A Planning Guide For Small and Medium Size Wood Products Companies:
The Keys to Success

I0470837

by

Jeffrey L. Howe
Research Assistant
Forest Products Management Development Institute
Department of Forest Products
University of Minnesota
2004 Folwell Avenue
St. Paul, MN 55108

Stephen M. Bratkovich
Forest Products Specialist
Northeastern Area
State & Private Forestry
USDA Forest Service
1992 Folwell Avenue
St. Paul, MN 55108

SEPTEMBER 1995

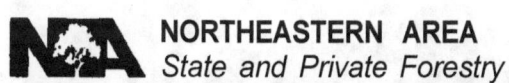

NORTHEASTERN AREA
State and Private Forestry

Management Development Institute
Department of Forest Products
University of Minnesota

Acknowledgments

The authors would like to thank the following individuals for their helpful comments and suggestions on this publication: Jim Bowyer, Harlan Petersen, Kevin Powell, and Mary Ferguson, University of Minnesota, Department of Forest Products; Dave Bengston, USDA Forest Service, North Central Forest Experiment Station; Bob Romig, Ohio State University; and Dentley Haugesag, Minnesota Department of Trade and Economic Development. Thanks is also extended to Gary Mitchell who developed the illustrations for the publication.

Table of Contents

Preface

The premise for writing this guide came from research studies in Maine and Minnesota which focused on success characteristics of small and medium size wood products companies. This guide was a cooperative project between the USDA Forest Service, State and Private Forestry, Northeastern Area and the University of Minnesota, Department of Forest Products, Forest Products Management Development Institute.

Printed on
Recycled Paper

The KEYS to Success!

Part I. Introduction

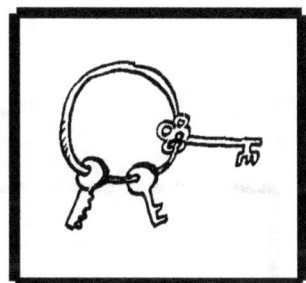

Today, wood products companies across North America are facing competitive pressure from numerous sources. Traditional products are being manufactured in new regions (e.g. developing nations), substitute products are being developed by competing industries (e.g. steel), and there is a strain on bottom lines brought about by greater restriction of natural resources and the general rising cost of doing business. All this is stretching the abilities of the wood products executive to the limit. The times are changing, and they're changing quickly. So what can be done to improve the chances of success and maximize the firm's capabilities?

Research has shown that a <u>formal planning process</u> is a key element in the success of manufacturing companies, especially with regard to developing new products and new markets. Recent research[1] on wood products companies in Maine and Minnesota demonstrated that wood products firms recognized as industry role models were significantly more likely to have a formal planning process than the industry as a whole. Formal, written plans allow firms to create strategies and programs to implement their ideas, and help them to measure and control their results. Formal planning is not a guarantee of success, but it is a process that facilitates the bringing of skills and abilities to bear on the proper issue, at the proper time.

We have written this guide with the objective of providing you, the wood products executive, with a tool to make your planning process a little simpler, quicker, and more complete. We can't tell you how to run your business. You undoubtedly already have the knowledge, training, and experience that is needed to direct and lead your firm to a successful and profitable future. But, what we do in this guide is to provide a framework for organizing your thoughts, mobilizing skills, and formalizing plans so that your ability to implement, measure, and control your business activities is maximized.

> *Research Shows:*
>
> *"Excellent" Small and Medium Size Wood Products Companies have Formal, Written Strategic, and Marketing Plans!!*

About the "Guide"

We have attempted to make this guide as straightforward, complete and "user friendly" as possible. To that end we have tried to eliminate jargon or "techno-speak" wherever possible. However, because there is always some terminology that is new to some people, or used in a new way, we have briefly discussed or defined certain terms within the text of this guide.

The first two sections of this booklet, **Creating a Strategic Plan,** and **Creating a Marketing Plan** develop plans that are designed for use "internally" by company management and staff. The strategic

1

planning process is presented first, as it is the broadest in scope, and because its' processes are fundamental to the other two types of plans discussed herein. Marketing planning is presented second, as it is often a logical second step to the strategic plan, and because the two processes have much in common.

Perhaps the first time that many firms realize a need for planning occurs when they seek financing. Both banks and investors require business plans before they proceed. For that reason, and for the sake of completeness, a separate section **"Writing a Business Plan"** has been included. The plan resulting from this process has a slightly different focus than either strategic or marketing plans, since it is usually directed at someone outside the firm.

We have organized this guide so that your firm can utilize any of the three different sections independently, as your needs require. At the same time we have attempted to limit redundancy. We ask readers to consider this guide in the same manner as they consider their plans. That is, both your plans and this guide are dynamic pieces of work: never perfect and never complete. In fact, at the end of each section we provide a "notepad" where you can add your own thoughts and suggestions, so that this guide can be continually improved, by you, for you.

In the end, we suggest that your plans not be set in stone, but also that your goals not be adjusted to meet your achievements. Use your plans to set your sights high, and to manage your activities to meet those objectives.

Remember, good management requires analysis, planning, implementation, and

control. Formal plans are the KEYS to making it all work right.

Follow Me!

The KEYS To Success!

Part II. Creating a Strategic Plan

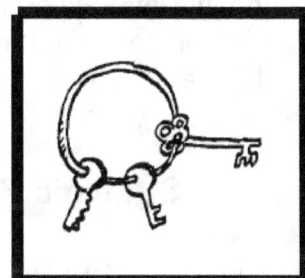

What is Strategic Planning?[2]

Strategic planning is concerned with shaping the nature and direction of your firm (*"What is our mission? Where are we going?"*) and developing broad strategies for accomplishing the mission (*"How do we get there?"*). Successful strategic planning involves making difficult choices, setting broad priorities, envisioning the firm's future, and developing procedures to achieve that future. The time-frame is long-term, 5-10 years or longer in some cases. Strategic planning is the responsibility of your firm's top management.

It is important to recognize the difference between the more modern "strategic" planning and the traditional long-range planning.

Strategic planning:

- focuses on identifying, managing, and resolving issues;

- emphasizes assessment of the environment outside and inside the firm; and

- is concerned with the "vision of success" of the firm and how to achieve it.

Long-range planning:

- focuses on achieving specific objectives of the firm;

- emphasizes management of the internal environment - the human, financial, and physical resources expected to be available to the firm; and

- tends to be based on historical projections and does not work well under changing external conditions.

Long-range planning focuses managers' attention on predicting rather than creating the future.

Where Are We Going?

Why Is Strategic Planning Needed?

Uncertainty, complexity, and change characterize the external environment in which businesses operate. Managers need to respond quickly and effectively to changing circumstances. Strategic planning helps to define an overall sense of direction and purpose for a firm, and thereby helps managers respond to change.

Several potential advantages or benefits of strategic planning may be identified, including:

Short-term benefits:

- raising awareness about the external environment

- improving dialogue among managers on strategy

- improving dialogue between managers and professional staff

- improving dialogue between the firm and its stakeholders

- building teamwork and planning expertise

Long-term benefits:

- providing direction and unity to the firm's efforts

- improving the firm's performance

- stimulating forward thinking in the firm, especially among top managers

The last point is perhaps the most important. Strategic planning is not an end in itself, but should help managers think and

act strategically. Successful businesses have always been guided by strategic thought and action, and a strategic planning process can help your firm develop this perspective.

A Strategic Planning Process

The approach used for strategic planning should be adapted to suit the nature and circumstances of the particular business or organization. Strategic planning in a large corporation will differ in certain ways from strategic planning in a small firm, e.g., in the size of the planning team, resources devoted to the planning effort, the amount of outside involvement. But the essence of strategic planning is a process of thinking that is largely independent of scale and is useful in all types of businesses.

How do you get started?

This section provides a step-by-step approach to strategic planning that can be adapted to the needs of a particular wood products firm, and implemented by current employees at reasonable cost and in a timely fashion (see figure in Appendix A). The planning process involves eight steps:

1. Initiating and agreeing on a strategic planning process
2. Identifying and clarifying values
3. Conducting a stakeholder analysis
4. Developing a mission statement
5. Assessing the external and internal environments
6. Identifying strategic issues
7. Developing strategies to manage strategic issues
8. Implementing and monitoring the strategic plan

4

The following sections describe these steps.

Step 1. Initiating and Agreeing on a Strategic Planning Process

(In the beginning...)

The first step in strategic planning is to reach initial agreement about the nature, purpose, and process of strategic planning. A planning committee should be formed to address the following important questions: Who should be involved in the effort (individuals and organizations)? Who will be on the strategic planning team? Who will oversee the effort? What are the potential benefits to the firm of strategic planning? What resources are needed to proceed with the effort? What are the desired outcomes? What specific steps should be followed? What should be the form and timing of reports?

Key decision makers should be included on your strategic planning team, and perhaps some representatives of important external "stakeholder" groups (e.g., representatives from raw material suppliers, customers, trade associations). On the other hand, you may decide not to involve external stakeholders initially - outside involvement may complicate the process. Regardless, your planning team should include individuals who tend to be innovative and insightful in strategic thinking and decision making.

The number of people directly involved in strategic planning will depend on the size of the company. In small firms, by necessity, the planning team may consist of a couple of individuals. In medium-size companies, two groups may be required to get the job done: a relatively large group to provide broad representation and

legitimization of the planning process, and a smaller committee that does most of the actual work and makes recommendations to the larger group.

Team Should Welcome Key Decision Makers!

Step 2. Identifying and Clarifying Values

(We believe in....)

All business decisions are based on values. A value can be defined as a conception of what is good or desirable, and serves as a guide to action. Many companies (perhaps even your firm) are not clear about their values. This lack of clarity can cause problems to emerge that could be largely avoided through a widely agreed upon set of underlying values. Without clarifying values early in the planning process, you will find it difficult or impossible to develop a usable strategic plan.

A big part in identifying and clarifying values is an examination of the personal values of the members of the planning team. It's quite important for the key decision makers in your company to be clear about their personal values and to recognize those differing values that exist among them. For example, an individual who values risk taking will view a company's future quite differently from a person who holds security as a high personal need. Planning team members with differing values should talk to each other and attempt to resolve the differences. Also, if any key decision makers in the company are not part of the planning team, then you also need to assess their values.

Once individual values have been worked through, the desired values of your company must be considered. Profit versus growth, being a good corporate citizen, being seen as a value-added company or a good place to work are examples of issues the planning team will likely discuss. The important issue to address is what values the planning team wants to have your company operate under in the future. As an example, these values may focus on performance ("excellence"), people ("employee pride and enthusiasm"), and process ("teamwork throughout the organization").

A company's values are often organized into a set of operating principles. For example, the Mars Corporation (candy manufacturer) operates on the following five principles:

1. Quality...The consumer is our boss, quality is our work, and value for money is our goal.

2. Responsibility...As individuals, we demand total responsibility from ourselves; as associates, we support the responsibilities of others.

3. Mutuality...A mutual benefit is a shared benefit; a shared benefit will endure.

4. Efficiency...We use resources to the fullest, waste nothing, and do only what we can do best.

5. Freedom...We need freedom to shape our future; we need profit to remain free.

The key message is that the strategic plan must be in harmony with your firm's values. If not, either the values or the plan must change.[3]

Step 3. Conducting a Stakeholder Analysis

(What do others think?...)

Stakeholders are defined as individuals, groups, and organizations who will likely be impacted by or who will likely be interested in your company's strategic plan and the planning process. Examples of stakeholders for a wood products industry could include employees, raw material suppliers (landowners, loggers, sawmillers, lumber brokers, etc.), customers, investors and lenders, debtors, interest groups (industry/trade associations, conservation groups), and other wood-using industries. Important employee groups should be explicitly identified as stakeholders.

A key to the success of any company is the satisfaction of key stakeholders. Any company that does not have a clear idea of who its stakeholders are, what they expect from the company, and how they judge the organization will have little chance of satisfying them. The stakeholder analysis can be structured around the following questions:

* *Who are my company's stakeholders?*

* *What do they want from our company?*

* *What criteria do the stakeholders use to evaluate our company?*

* *How is the company performing against those criteria?*

The first question can likely be answered through a brainstorming session of the strategic planning team. You may find it useful to rank the stakeholders according to their importance to the company.

The second and third questions can be approached in two ways. One is for the planning team to make informed judgements about the wants and evaluation criteria of stakeholders. The second approach is to ask stakeholders, through interviews or surveys, what their wants and criteria are. The first approach is much faster and avoids any problems with stakeholders not being completely honest. For example, a raw material supplier may be concerned primarily with the company's continued purchase of low-grade lumber at an above-average price, but would be unlikely to publicly make such a comment. However, a significant risk in not asking for input from stakeholders is that the planning team will incorrectly assume what the views of stakeholders are.

The fourth question to be answered in the stakeholder analysis concerns how well your company performs against the stake-

holders' criteria. To get useful discussion on this question, it may be necessary to indicate whether your firm's performance is poor, average, or excellent relative to the different criteria (e.g., quality of final product, disposal and/or use of residues, customer relations).

The completed stakeholder analysis should serve as a starting point for discussion of exactly how the various stakeholders influence your company and which are the most important stakeholders.

Step 4. Developing a Mission Statement

(We want to be...)

The "mission" is, simply, the basic purpose of your company. A well-developed mission statement can be a valuable management tool, providing future direction and a basis for decision-making. A mission statement should ideally serve as a guide to what management wants your company to be. It should remind and motivate managers and employees to identify with the goals and philosophy of your company. Consequently, all aspects of business operations should be focused on achievement of the mission. Mission statements should also fulfill an important public relations role, by concisely communicating to stakeholders what your company is all about.

The stakeholder analysis provides information that is useful in developing a mission statement, but more is needed. In establishing a mission statement, your company must establish its "scope" in, at least, three areas:

- First who is to be served and satisfied by the company?

  ("Who are our customers?")

- Second, <u>what</u> is to be satisfied?

  ("What customer needs should we satisfy?")

- Third, how are needs satisfied?

  ("How can we most effectively satisfy customer needs?")

Essentially, these three key questions can be rephrased as one:

"What is our reason for being?"

It is extremely important for your company to achieve a clear definition of the "reason for being." For example, there is considerable contrast between a small firm specializing in custom-made furniture and a large-scale ready-to-assemble furniture company. Both are wood products companies and both produce furniture, but both are very different in their "reason for being."

What Does A Mission Statement Look Like?

Consider the following mission statement:[4]

XYZ Woodworks Company is committed to excellence in the manufacture of solid wood components primarily for the kitchen and bath cabinet industry. Excellence means providing a consistently superior product - on quality, on quantity, on time. As we do this we serve the needs of our customers, our employees, our suppliers and our community

The *who* in this mission statement is the "kitchen and bath cabinet industry", i.e., the customer. The *what* is the "consistently superior product" (solid wood components). The *how* is "on quality, on quantity, on time". The *who, what* and *how* establish the "scope" of the XYZ Woodworks Company.

The example below also illustrates a company's scope by focusing on the three areas of *who, what* and *how:*

ABC Millwork Company

Our Mission,
as a millwork and hardware specialist,
is to be the standard of excellence
for our customers in the building industry.
We continually strive for excellence
by building and maintaining a team
of professionally skilled and
highly motivated employees
who are responsible for providing
quality products and services
to our successful customers.

Defining and/or clarifying your company's mission can be a soul-searching, demanding and time-consuming process. Different individuals may have conflicting views of what the company is about and should be about, Each member of the strategic planning team should answer the "who, what, and how" questions individually first, and then come together as a group for discussion. After the group discussion, an individual should be assigned to develop a draft mission statement. The draft mission statement should be discussed and modified as needed throughout the remainder of the strategic planning process.

Though the process of defining your firm's mission may seem like a lofty exercise for a small, or medium size company, it is not meant to be simply window-dressing. The process of planning team members wrestling with the mission statement may be as important as the statement itself.

Step 5. Assessing the External and Internal Environments

(What's up Doc?...)

A major purpose of strategic planning is to identify external threats and opportunities that may demand a response in the near future. The idea is to prepare your company to respond effectively before a crisis develops or an opportunity is lost. Assessing trends in the external environment is

Check Your Business Climate.

therefore an important part of strategic planning. What are the recent issues and emerging trends affecting your company? These include environmental, technological, political, economic, and social trends

and issues that may be local, national, or worldwide in scope. For example, reduced timber harvests from public forests in the western U.S., boycotts of tropical timber products, or increased regulation of logging on private lands can all have direct impacts (both negative and positive) on wood products companies throughout the country.

Some large corporations use formal "scanning" procedures for assessing external environments. However, for small to medium size firms, elaborate and demanding procedures are generally less desirable than simple and practical methods. Many companies rely on the knowledge of members of the strategic planning team and use group discussions to identify external threats and opportunities and assess their significance to the firm. Other approaches you might want to consider include organizing workshops or discussion groups involving stakeholders to identify major issues, or using various survey techniques.

The internal environment should also be assessed to identify strengths and weaknesses that help or hinder your company in carrying out its mission. Broad categories of internal strengths and weaknesses include:

- resources available to your company (such as technical and support personnel, equipment, facilities, computer resources, or ownership of forestland.)

- company structure (organization of the firm, decision-making, chain-of-command, allocation of resources.)

- company performance (products and services and how these impact customers)

Using these categories, your planning team should develop a list of major internal strengths and weaknesses of the company. This list, along with a list of external opportunities and threats, should then be discussed and analyzed. Scanning and assessing the external and internal environments should be a continual activity so that relevant information is always available to your firm's key decision makers.

Step 6. Identifying Strategic Issues

(What's coming at us?...)

The previous five steps of the strategic planning process lead to the identification of strategic issues. A strategic issue is a fundamental policy choice facing your company. Strategic issues call for a reexamination of your company's values, mission, and the kinds, level, and mix of products and services provided. Strategic issues usually arise when:

- external events beyond the control of your company, make or will make it difficult to accomplish objectives given the resources available;

- choices for achieving company objectives change, or are-expected to change (e.g., changes in raw materials, technology, financing, management); or

- new opportunities arise.

Examples of strategic issues that a wood products company might face include a dwindling supply of quality logs or lumber (raw material), substitution of non-wood products for traditional wood products (steel studs vs. 2x4's), increasing conflicts among groups that utilize forests, and a pending new discovery relating to wood

residue utilization.

The process of identifying strategic issues involves first reviewing the value, mission, external threats and opportunities, and internal strengths and weaknesses. Each member of your planning team should be asked to individually identify strategic issues by answering three questions for each issue:

> *Research Shows:*
>
> *Excellent Wood Products Firms Rank the Availability of Wood as a Significant Limit to Their Growth!*

- What is the issue? (describe in two or three sentences)

- What factors make the issue a fundamental policy question? (How does the issue affect values, mission, internal strengths and weaknesses, etc.?)

- What are the consequences of not addressing the issue? If there are no consequences, it is not a strategic issue. If the company will be significantly affected by failure to address an issue or will miss an important opportunity, the issue is highly strategic and should receive high priority.

Planning team members will need time to reflect on these questions, and at least a week should be devoted to individual identification of strategic issues. The entire planning team then convenes and tentatively agrees on what the issues are. Each issue should be summarized on a single page, addressing each of the three questions posed above. Strategic issues are then prioritized to aid in developing strategies to deal with the issues.

Step 7. Developing Strategies to Manage Strategic Issues

(Should we do this or that?...)

A five-part process for developing strategies to manage strategic issues is recommended. For each issue that your planning team has identified, the following questions should be addressed:

1. What are the practical alternatives your company might pursue to address a particular strategic issue?

2. What are the barriers to realization of these alternatives?

3. What specific actions might be pursued to achieve the alternatives directly or to overcome the barriers?

4. What actions must be taken within the next year?

5. What specific steps must be taken within the next six months, and who is responsible?

The purpose of these questions is to clarify what has to be done and who has to do what to deal with each strategic issue. For example, suppose a strategic issue facing a wood products firm (posed as a question that the company can address) is: <u>How can we best recruit and retain highly talented and qualified employees?</u> Practical alternatives to address this particular issue might include:

- Better anticipate shortages of trained personnel.

- Simplify hiring practices.

- Improve the system of rewards and incentives to increase retention of employees.

- Develop and maintain close ties with universities, colleges, and technical schools to identify potential employees for recruitment.

Using the last alternative as an example, potential barriers to realizing this alternative might include:

- Proximity of company to universities, colleges, and technical schools.

- Lack of knowledge of specific programs and degree offerings.

- Lack of an ongoing employee recruitment program.

11

Proposals to achieve the alternative directly (develop and maintain close ties with universities, colleges, and technical schools to identify potential employees for recruitment) or to overcome the barriers might include:

- Develop a student internship program with a university, college, or technical school.

- Participate in student job fairs and related employment-seeking activities.

- Become active in steering committees and other volunteer groups that deal with curriculum development, course offerings, and student recruitment.

The last two questions of the five-part process involve identifying the specific actions that need to be undertaken and assigning responsibility for carrying out the strategy to an individual or committee. However, your planning team may address only the first question - identifying practical alternatives to deal with a strategic issue - and a key individual would then be assigned to follow up on one or more of the alternatives as part of the implementation of the strategic plan.

Step 8. Implementing and Monitoring the Strategic Plan

(Here we go! . . .)

The final written strategic plan should be a summary of your planning team's efforts, usually limited to 10 to 15 pages. The simplest form for a written strategic plan consists of the final versions of some of the parts completed by the planning team, such as:

- Mission statement

- Company values (formal and informal)

- External opportunities and threats (factors that might affect the direction of future activities)

- Internal strengths and weaknesses

- Strategic issues facing the company

- Strategies to manage the issues

- Plans for implementation

A key person in your firm should be selected to prepare the first draft of the written strategic plan. The draft is then reviewed and modified by other members of your planning team, by key decision makers, and possibly by external stakeholders. After a final review by internal and external reviewers, including employees, the plan will be ready for formal adoption and implementation.

Key Person Prepares 1st Draft!

How will this Affect Your company?

The true test for any strategic planning process is the extent to which it affects the activities of the company and the behavior of employees. The purpose of strategic planning is to develop a better road map to guide the company. Unless this road map actually guides decisions and actions, the planning process is nothing more than an academic exercise. Implementation of the strategies developed, however, does not follow automatically. Change will be threatening to some and almost always faces resistance. Resistance to the implementation of strategies may take the form of procrastination, "paralysis by analysis", lack of implementation follow up, or even outright opposition.

The role of top management in implementing your firm's strategic plan cannot be overemphasized. The "Boss" should be closely involved in the process from the outset and must be totally committed to the strategic plan and lead the support. All managers (or whatever you call them in your firm) need to be committed to the plan and use it to guide decision making, particularly in developing annual objectives and budgets. Management must communicate the plan and its rationale to all employees, and employees must be intimately involved in the planning process, rather than being presented with a completed plan.

Periodic review and updating of your strategic plan and monitoring of its implementation are vital. Every six months the specific steps to implement the plan should be reviewed. Every year or two the strategies to deal with strategic issues should be reviewed and progress evaluated. And every three or five years the entire strategic plan should be reviewed by the strategic planning team and modified as needed. Many companies review and update their strategic plans yearly, before planning and budgeting for the coming year.

Top Management Leads Support!

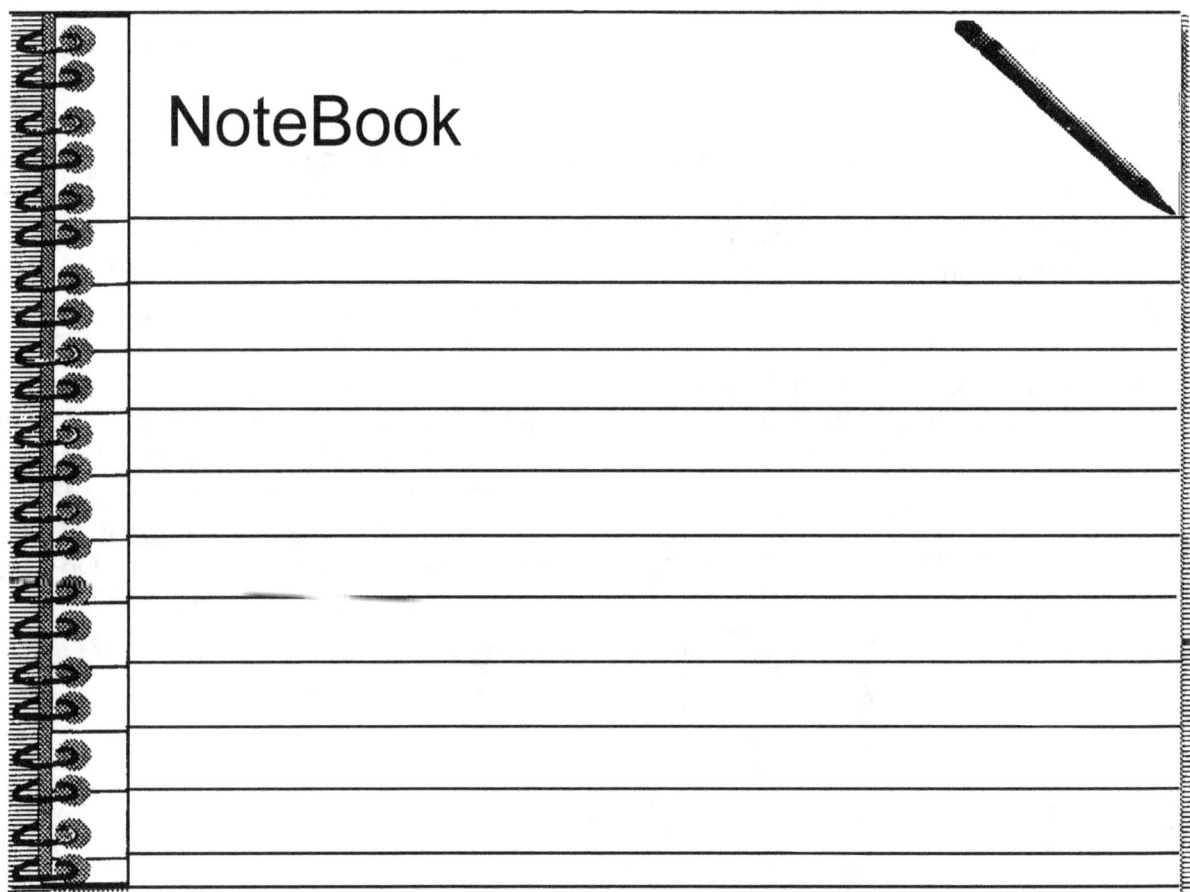

The KEYS to Success!

Part III. Creating a Marketing Plan:

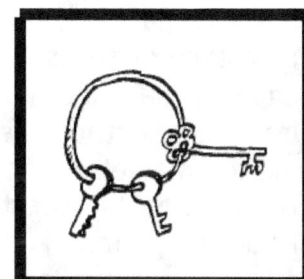

What is Marketing?

By one definition, Marketing is simply: "a social and managerial process by which individuals and groups obtain what they need and want through creating, offering and exchanging products of value with others."[5]

A more academic approach suggests: "Marketing consists of individual and organizational activities that facilitate and expedite satisfying exchange relationships in a dynamic environment through the creation, distribution, promotion, and pricing of goods, services, and ideas."[6]

Whichever definition you prefer, recognition of the customer as the focus of the company's activities is the first step to developing a marketing orientation.

What is a Marketing Orientation?

A marketing orientation is the focusing of ALL company activities toward meeting the wants and needs of the customer. Excellent firms emphasize a marketing orientation as part of their corporate culture. This recognizes the integrated nature of product design, development, quality control efforts, sales, after-sale service, and "marketing."

Why is a Marketing Orientation Important?

Back in 1960, Harvard Professor Theodore Levitt wrote his famous article "Marketing Myopia,"[7] in which he discusses how management's failure to define their product broadly enough can lead to a decline in revenues and/or a susceptibility to competition from new industries or new entrants.

The railroad industry has yielded a classic demonstration of marketing myopia.[8] For years it was readily apparent that the primary role of the railroads was to transport both passengers and freight. *Yet, as the need for passenger and freight transportation grew significantly throughout this century railroads became less and less profitable!* Although there were a number of factors involved, most of the problems arose because customer's needs began to be filled better by other means (for example: cars, trucks, airplanes, and even telephones).

Railroads assumed themselves to be in the railroad business, rather than the transportation business. They were product-oriented, rather than customer-oriented. As a result, many railroad companies went out of business and the industry lost its dominant position as a transporter of people and freight. However, since the early 1980's railroads have been redefining themselves; adding trucking services, freight forwarding service, even air service. Railroads are regaining a customer focus, and have begun a process aimed at reestablishing lost markets.

Wood products firms are redefining themselves as well. For example: In response to the steel industry's aggressive marketing efforts to capture an increased share of the residential framing market, some wood

15

products firms have recognized that the steel industry has identified certain needs not currently being adequately met (such as a desire for more dimensionally stable framing). Instead of simply arguing it out in the promotional literature, these firms have sought solutions in the products themselves. The development of finger-jointed materials, composite products such as I-joists, framing with a lower than normal moisture content, and environmentally certified wood are all new solutions to the changing needs and demands of customers. In this instance the manufacturers have redefined their business from being "lumber" producers to the broader "wood products" manufacturer.

It should also be noted that changes affecting businesses do not have to be either subtle or sudden to take a firm by surprise. Consider the following quote (In reference to the increasing impact of large regional supermarkets):

"...it's hard to believe that people will drive for miles to shop for foods and sacrifice the personal service........to which Mrs. Consumer is accustomed."[9]

Does this statement sound familiar? Although this statement is very applicable today, it is actually from a grocery executive in the 1930's. Haven't we recently heard similar comments about the large Home Centers? The underlying trend has not changed, only the definition. The gist of all this is that a customer orientation is critical to your firms longevity, and this is exactly what marketing is all about.

What is a Marketing Plan?

A marketing plan is a brief written report that summarizes the details of your firms marketing activities for a given period in the future; usually the next year. It is the "central instrument for directing and coordinating the marketing effort."[10] As such, It is the blueprint for the construction of your marketing programs.

The marketing plan differs from the strategic plan, in that it focuses more narrowly on a specific product and/or market. It provides a detailed framework for the marketing process, including an entire set of specific activities to be performed to meet management's strategic goals and objectives. The marketing plan is limited in that it primarily deals with strategy as it relates to the company's marketing mix.

Marketing Plan Directs Marketing Efforts!

What is My Company's "Marketing Mix?"

The marketing mix is the combination of Product attributes, Pricing strategies, Promotional strategies and distribution methods that are developed to best meet the needs of both your customer and your firm. These factors are usually referred to as the 4-P's of marketing (distribution is referred to as "Place").

How are Marketing Plans Developed?

Marketing plans are developed in a two stage process. In Stage 1 information is gathered and analyzed. In Stage 2, the information is reorganized and summarized in a brief written plan. To make it easier to understand and discuss, this marketing process can be subdivided into the following activities:"

Stage 1:

1. Analyzing Marketing Opportunities
2. Evaluating Markets
3. Designing Marketing Strategies
4. Planning Marketing Programs
5. Organizing, Implementing, and Controlling Marketing Effort.

Stage 2:

1. Developing the Written Marketing Plan

The importance of step 1) Analyzing Marketing Opportunities and step 2) Evaluating Markets, in stage 1, cannot be overemphasized! Too often wood products are brought to market for the "wrong" reasons, i.e., firm owners really "like wood," or they "like machining," or they simply have access to both wood and machines to make a product. However, by placing an analysis of the markets

> *Excellent Firms Place High Importance on Pre-Development Activities!*

early in the planning process, your firm recognizes that it must be "market driven" rather than "product driven." This recognition is crucial to the success of a product.

Let's take a brief look at all of the steps in more detail.

Stage 1:
Step 1. Analyzing Opportunities

(What could we do ? . . .)

To some, analyzing their opportunities is the most difficult activity to visualize. The key to analyzing marketing opportunities often rests on the quality of your firm's marketing information system. Your ability to meet your customer's expectations is dictated by how well you research their wants and needs, their location(s), buying habits, inventory methods, and so on. There are two common approaches to achieving this. One involves a review of written information (sales reports, old marketing research, and a review of library information are examples) and the second involves personal interviews with customers and industry "experts." Usually firms are already using this latter approach via their sales staff. Sales personnel are often the eyes and ears of the company, but too often what they know is inaccessible to all but themselves.

The crucial part of the process of analyzing opportunities occurs when the relevant information is formally **written down** and processed in a manner that facilitates its use by management and sales personnel alike.

Smart People Have Written Plans!

Good marketing opportunities can involve new services to existing customers, new products and/or new markets (new customers!). Research has shown that there are a number of factors that have a significant impact on the success rate of new products and new markets. Key among these factors are:[12]

A. New markets & products should be selected that fit well with the firm's existing expertise.

For Example:

1) Choose new products that have technological needs similar to existing capa-bilities (e.g. is special training involved?), have handling requirements concurrent with existing capabilities and experience, and require a similar company attitude.

More specifically, a manufacturer of pallet stock may want to take the intermediate step of making dimension parts, before leaping into cabinets or furniture manufacture, simply because of the dramatic change in company "style" that might be necessary to produce cabinets or furniture.

2) Similarly, choose new markets that are similar to existing ones. Successful firms tend to have a specialty, whether it is supplying mobile home manufacturers, furniture manufacturers, retail yards or wholesale distributors. In marketing we call this your niche (it doesn't mean it's small!). Evaluate the existing customers that you are most successful with, and then search out their counterparts in other areas/regions. Often the skills that made you successful with them, plus the experience you have gained in serving them well, makes your development of new, similar accounts much easier.

B. Look for products and markets where what you supply has a clear differential advantage in the eyes of the customer:

Research Shows:

Excellent Firms Believe that Distinctive Product Features are VERY Important!

18

For Example:

1) Does the product or service you provide clearly meet the customer's needs better than your competitors?

2) Does it have certain unique qualities that allow the customer to distinguish your product from all the others out there?

3) Does it help the customer reduce net costs? Or do their job better?

4) Is the product new? Creative? Innovative? A new application of an existing product?

C. *Make your choice of a new product and/or new market based on sound marketing research and a formal marketing process. Basically, do your homework first; put it all in writing, and evaluate it based on realistic goals and objectives.*

For Example:

1) Have you gone to the library to see if there are applications for your product you haven't thought of previously?

2) Have you talked to industry experts to get their opinions of what to do, or how to go about it?

3) Have you discussed this with your entire staff (including clerical, sales, and delivery people)?

4) Have you looked at all your historical data (sales & profit figures, inventory histories, etc.) to anticipate trends?

5) Are your promotional expenditures in line with your sales objectives?

6) Have you analyzed and formulated your complete marketing mix for the new endeavor?

Step 2. Evaluating Your Markets

(Where can we go?...)

So now you've done the background work, you've selected a product(s) to go forward with and you have some idea of the type of customer you want to go after. So how do you pick exactly which customer to go after? How much are you going to sell them? Are they worth the effort? How do you forecast sales to this new target? and, How should you position your "offer"?

The basis of evaluating markets lies in the segmentation, targeting, and positioning of your products.

Segmentation

First, we need to describe a Market. Basically, a market has been defined as an aggregate (group) of people who, as individuals or as organizations, have needs for products and who have the ability, willingness, and authority to purchase such products.

Segmentation is simply the process by which the total market is divided into groups consisting of those who have relatively similar product needs, in order that company resources may be more accurately focused to meet those needs.

Most wood products firms are familiar with the basics of segmenting their markets, so we won't dwell on it here. Suffice it to say that some of the more common approaches to segmentation involve:

19

- The use of demographic variables so as to separate customers based on size of company, credit rating, time in business and volume of product they use. This could also mean separating customers based on demographics of the main buyer at the customer firm, such as age, gender, education, etc.

- The use of geographical variables so as to separate customers based primarily on location, but also based on climate, terrain, natural resources and cultural values. This approach is very common in the wood products industry - to some extent as a result of freight costs, and to some extent due to tradition.

- The use of psychographic variables so as to separate customers based on personality characteristics. For example, what are the habits, interests (do they golf?), and life-styles of both the individual buyer and the company

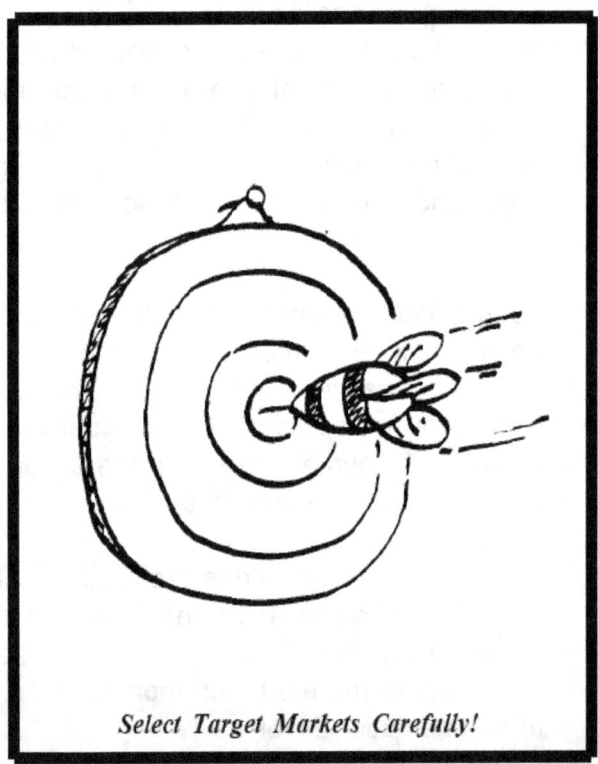

Select Target Markets Carefully!

they represent. Does the company put out bid lists? Do they like to visit the mill first? Do they need written or verbal confirmation? What is their corporate "culture"? These are the activities, interests, and opinions of both the individuals and firms themselves.

Selecting Targets

The selection of a target market is simply the matter of choosing the segment(s) of the market whose needs you are best able to meet, and which also meet the requirements you have selected as being optimum for a "good" customer. This is primarily a matter of matching your strengths to customer needs, in areas where your weaknesses are minimized. As part of target selection, a good idea is to make some form of forecast of the sales potential for the targets you have selected, to be sure you have chosen the ones most worth the effort.

Forecasting Demand for Your Targets

Once you have a few targets defined, there are a number of ways of forecasting demand for your product. Some are quite simple; some extremely complex. Here are five common ways.

1. Sales Force Estimates - Here you ask your sales personnel to estimate expected sales by customer, and by product. Management may want to occasionally adjust such estimates based on their greater knowledge of market conditions. But these kinds of estimates tend to gain accuracy over time (with annual

use), and have the benefit of being extremely detailed. In addition, it is often suggested that sales representatives have greater confidence in their quotas, and more incentive to achieve them, if they participate in the forecasting process.

2. Buyer Interviewing - Here your sales force, or an outside agency, goes through the process of asking buyers, or potential buyers, detailed questions about their anticipated needs. The difficulty is that buying habits can be erratic, or buyers simply don't know what to expect. However, for some industries and some firms, purchasing behavior can be projected out at least 3 to 6 months. You just have to ask!

3. Expert Opinion - We're back to those experts again. This should not be surprising, as consulting with industry experts is probably the approach used most commonly by wood products firms today. With this approach, you simply ask the people you respect most what they expect. The difficulty here is finding those with the knowledge specific to your new markets, and to do this you may have to use referrals. Ask at trade shows, association meetings etc. to find out who is familiar with your new market, and call or go see them. However, you might be surprised by how many of your suppliers and other existing contacts are already familiar with what is new territory to you. Use these resources.

4. Time-Series Analysis - Here you evaluate data on historical sales of products similar to yours, or ones that fulfill a similar function, for the market you are interested in. There is a sur-

prising amount of data available in libraries and at universities that provide a fairly complete historical record. From this, and any data that you can obtain directly (such as looking at your own historical trends) you attempt to create a portrait of the activities for the coming year(s).

5. Market-Test Method - This method is especially desirable for a completely new territory or a completely new product. There are numerous marketing texts on this process, but the gist of the story is that you try out your whole marketing mix on a very small scale (but in proportion), to get an idea of what to expect.... before you launch into full scale production.

It is very common for wood products companies to try pieces of this last approach, in that they will make a few samples of a new product and sell them to a couple of close customers for feedback. The difference here with the market-test method is that you *simultaneously test not only your product, but the entire marketing mix* of price, product, promotion, and distribution. From a small test over a short time period you develop some estimate of the sales volume you might expect from the total market, and this process also provides feedback on the effectiveness of your marketing mix.

At this point you've segmented your customers into groups and have developed a forecast of your sales volume for certain targets. The next step is to decide how to position your product for final promotion. Selection of your product positioning approach occurs as part of designing your overall marketing strategies.

Step 3. Designing Strategies

(How should we do this?...)

What are Marketing Strategies?

Marketing strategies define the approach by which a firm plans to achieve its specific goals and objectives for a specific product in its target market(s). These strategies tend to be broad in scope.

Marketing strategies are designed with the objective of differentiating and positioning your product in a target market. With each step in the process you are getting more and more focused and more and more precise in defining how your firm and customer will interact. In designing your marketing strategies it is necessary to take into account what the situation is, and determine what should be done to achieve success.

What do we mean by "product positioning?"

The positioning of your product is a critical part of the marketing strategy process, and specifically involves those activities by which you create the image of both your company and product in the eyes of the customer.

For example, you won't get a premium for your product if your image is one of a low-cost discount producer, and both your price and your promotion focus just on your product's price.

Are there any tools that help you position your product correctly?

To position your products profitably, you need to know exactly what things are important to your customer, and how well your firm performs to meet these customer needs. One very handy tool available to help visualize this is the Importance-Performance grid (see Figure 1).

To develop this grid you first must survey your customers and have them rank the attributes in question. For example, how important is on-time delivery to your customers, and how well are you performing *in their eyes?* Usually these attributes are ranked from 1 to 5, with 5 the highest. You would want separate questions asking "How important is this attribute?" and "How well does my company provide it?" A couple of sample questions that would go with the grid are given below.

a. How Important is it for your company to receive same-day delivery on the product(s) you order? (circle one)

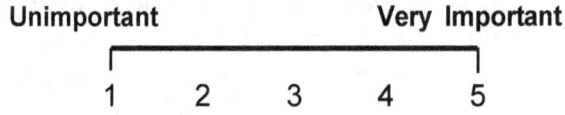

b. How effective is our firm at providing the type of delivery service your firm desires? (circle one)

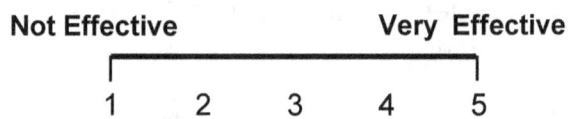

In this case "effectiveness" is your performance rating for this attribute! You would then ask similar questions for a number of different products/services that you provide, remembering that each attribute requires both an importance and a performance type question. These questions can be easily administered by your sales staff, either over the phone, in person, or through the mail. They don't have to be presented in a structured survey, but can be cloaked in normal conversation. As you gather input on a number of different attributes you begin to paint a very clear picture of the customer's specific wants and needs. Written questionnaires can even gain insight into some of the customer's unspoken and/or subconscious wants/needs.

Once you have gathered your customer's responses, there are three ways to plot them on the grid:

1. You plot a separate grid for each <u>customer</u>. On this grid you plot your customer's response for every attribute. This allows you to compare how well you meet the specific needs of this customer for the different attributes.

2. You plot a separate grid for each attribute. On this grid you plot the responses of the different customers for one attribute only! This allows you to compare how well you meet the needs of different customers for this attribute.

3. You plot one grid with the <u>average of all customers</u> responses for each attribute. This allows you to compare how well you meet the needs of customers, in general, for each attribute.

You interpret the grid depending on which quadrant your result falls into. Basically:

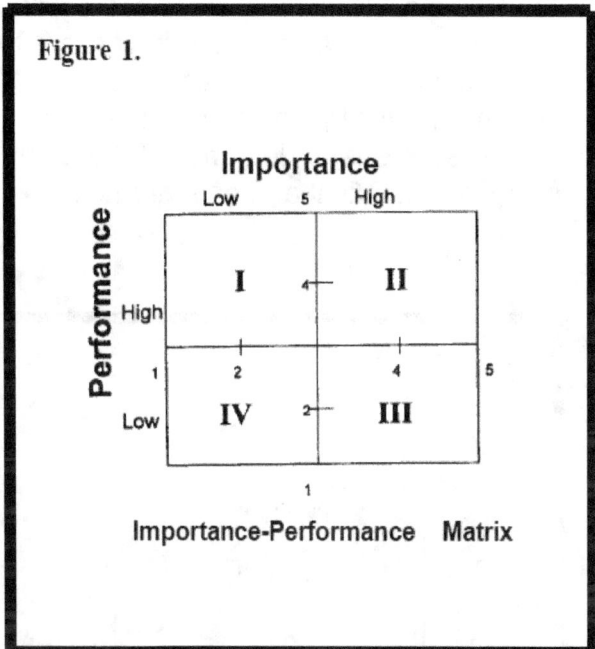

Figure 1.

Importance-Performance Matrix

Quadrant I: If you have high performance and low importance, you may want to de-emphasize this attribute, and it certainly won't help in your sales pitch. Keep up the good work, but don't brag about it.

Quadrant II: If you have high performance and high importance, you want to broadcast this at every opportunity. Emphasize this to your customers in every way you can and consider opportunities for premium pricing.

Quadrant III: If you have low performance and high importance you need to work very hard to improve your efforts. You may need further, more specific information from your customer to learn how you can improve in their eyes! But, if it's important to them, its worth the effort.

Quadrant IV: If you have low performance and low importance, you may not need to worry about it. Don't waste a lot of time and money improving areas your customer could care less about.

The Product-Positioning Map (see figure 2) is another tool that is similar in concept to the Importance-Performance Grid, but portrays a slightly different picture. The objective here is to determine if there is a niche you can fill that others are not.

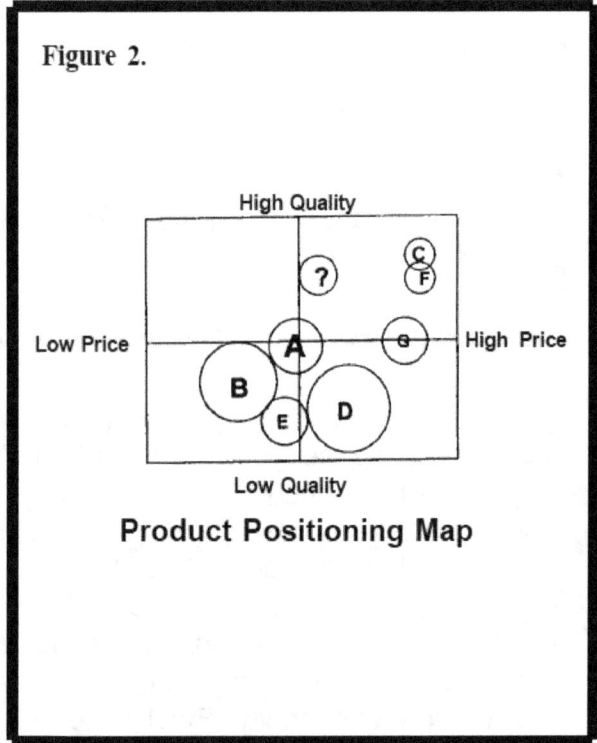

Figure 2.

Product Positioning Map

The size of the circles on the Map indicate the relative size of the different firms in the marketplace. The location of the circles indicates the price/quality relationship of your product with that of your competitors. In our example, there appears to be an opportunity for a firm to produce a high quality, moderately priced product without much competition (see "?"). However, an important point here is that the firm must be realistic about how they evaluate their own quality. Quality is always judged from the viewpoint of the customer. If a customer can't discern a difference, there isn't one!

In developing your company's strategies you must also take into consideration existing products in existing markets, and their position in the product life cycle. That is, are they still in the growth stage, the maturity stage, or is the market for these products generally in decline? Each of these stages require different strategies.

Finally, you would want to consider your firm's role in the marketplace. Are you a market leader? Are you challenging someone else's lead? Are you simply a follower of the "big boys?" Or are you carving out your own niche? Again, each of these approaches requires careful consideration, recognition of their differences, and each approach requires the use of a different strategy for successful implementation.

Step 4. Planning Marketing Programs

(Who does what?...)

Now we are getting into the 'nitty gritty" of the marketing planning process. Marketing programs define how things are to be achieved, how they are to be paid for, and how company resources (personnel, equipment, etc.) are to be allocated to achieve the firm's goals and objectives. It is here that the company "fleshes out" its marketing mix.

Product: Exactly what is your fully "augmented" product? This means, what is the complete package that your customer is paying for? What service level, warranty, accessories, packaging, credit, etc. are you providing for the price?

Price: How are you pricing your product, and is this going to change for your existing project? Do you have some advantage (patent, location, high entry costs for competitors, etc.) that allows you to realize higher than average profits from your market or with your new product? Are you trying to quickly "penetrate" a new market and are thus keeping margins tight to get as many new customers as possible? Maybe you are somewhere in between these two strategies.

Promotion: How are you communicating your benefits to your customers? Are you going to advertise, and if so, in what media? What is your message? Do you have sales brochures? Who does the sales? What about public relations?

Place: How are you going to distribute the product? Will it be sold in single units, in volume, and how, when, and where will it be stored? What channels will you go through to get to the final customer? Do you go direct to retailers, use wholesalers, or skip them all and sell only to industrial accounts?

All these questions (and more) must be answered in designing your marketing programs, and at this point you must begin to assign a budget to each of the individual activities. You are finally hammering out the details of how things are going to get done.

Step 5. Organizing, Implementing and Controlling the Marketing Effort

(Is it working?...)

The final part of Stage 1 in the marketing planning process involves decisions concerning who is going to do what in your plan, so that it is implemented in the most effective manner. It also involves development of control systems to be sure results are continuously monitored, activities modified as necessary, and that the firm adapts to new situations. Continual feedback must be obtained and reviewed to assure that the firm is adjusting to both macro- and microenvironmental changes.

Stage 2:

Step 1. Developing The Written Marketing Plan

(Let's put it on paper!...)

The formal marketing plan is the brief written report (10-15 pages) that develops from the above process, and generally includes the following sections:

A. Executive Summary
B. Current Marketing Situation
C. Opportunity Analysis
D. Company Goals & Objectives
E. Marketing Strategy(s)
F. Specific Action Steps
G. Projections
H. Feedback & Control Systems

Keep Marketing Plans Concise!

A. Executive Summary

Your marketing plan should begin with a short summary of your major goals and recommendations. This should be about one page in length, and is simply provided to enable your readers to quickly understand the basics of the plan. An abbreviated sample might look something like this: '

The 1995 marketing plan is designed to increase WoodCo's sales of custom cabinets by 20 percent to $1.5 million. The profit objective is set at $165 thousand. It is anticipated that these increases can be met through the addition of one salesperson to the Wisconsin territory. A slight decrease in gross margin (from 12 to 11 percent) is anticipated due to nonrecurring costs associated with the hiring and training of the new sales representative. The marketing budget is set at $288 thousand, an increase of 25 percent over last year......

B. Current Marketing Situation

In this section you want to briefly describe what are often called the 3-C's of marketing... Customers, Company, and Competition. You want to include data on the markets, product and distribution activity, and a general environmental overview.

<u>Customers:</u> You want to include data on the market in general for your products and for the customers in your different target markets. Here you use data gathered earlier in the process to discuss issues like needs, buyer behavior, and customer trends. You should discuss the size and growth of the market.

Examples:

1) *The market for wooden outdoor playground equipment has been growing at a IO percent annual rate for the last 4 years.*

2) *Purchase of wooden playground equipment by municipalities, school systems, and day care enterprises has increased by over 50 percent in the last 2 years.*

3) *The National Association of Playground Builders anticipates that total consumption of outdoor playground equipment will exceed $200 million by 1998, and that over 60 percent of sets sold will utilize wood as the primary structural component.*

<u>Company:</u> Here you want to get into the specifics of the situation related to your products, such as sales, prices, and contribution margins[13]. Again, you want to show these for each of your major products for the past few years. This is *usually given in tabular form,* but would include some description/explanation of the trends and impacts on company profits. In this section,

you should also briefly discuss the distribution of your products, including the number of units sold in each channel.

Examples:

1) *The average retail price for Wood-N-Stuff chairs has increased by at least 10 percent in each of the last three years. As a result, chairs have the highest contribution margin of any Wood-N-Stuff product.*

2) *Competition from imported products has made the market for dining room tables very competitive, reducing average prices for this product line in each of the past two years by an average of 7 percent. Tables no longer contribute directly to fixed costs, but are an integral part of dining room sets which have an average contribution margin. . . . .*

3) *Countertops are sold through a variety of channels, including retail home centers, contractor yards, specialty kitchen/bath stores, and direct to contractors. Table XYZ shows the volume of countertops by product line sold through each of these channels by Cabinets-R-Us. It can be seen that 35 percent of all countertops were sold direct, 20 percent through home centers and contractor yards, and 45 percent through specialty stores.*

<u>Competitors</u>: In this section you should identify the major competitors and their products. They should be described in terms of their size, market share, product quality, marketing strategy, distribution methods, and any trends that are apparent, You should review this for each of your major competitors.

Example:

1) *West Lumber Co. is our primary competitor in the Norway Pine board market in central Iowa. They market a complete line of pine boards primarily for the residential housing market. They sell directly to the home centers in truckload and half truckload quantities. They are not known for the highest quality, but their service is excellent, their pricing is very competitive, and they have captured approximately 25 percent share of the market for #2 and #3 common in this state.*

C. Opportunity Analysis

Given the situation you have described above, what do you see as significant opportunities (or threats) that should be addressed? This is often described as the SWOT Analysis, or Strengths, Weaknesses, Opportunities, and Threats. These are usually separated into two separate categories for outside factors (Opportunities and Threats) and inside factors (Strengths and Weaknesses).

Your report should include several brief statements that describe the opportunities and threats facing the business and suggest, or imply, some action steps that might be taken.

Examples:

Opportunities

1) *Customers are showing an increased concern about the environmental impacts of building material choices. "Wood-N-Stuff" should consider use of wood only from sustainably managed forests.*

2) *Repair and Remodeling (R&R) is becoming an increasing portion of the total consumption of cabinets and millwork. "Cabinets-R-Us" should develop a line of cabinet doors,*

27

drawer fronts, and frames that can be used to upgrade existing cabinets in place, thereby avoiding the expense of removing and replacing entire "kitchens."

Threats

1) *The steel industry has developed a major national advertising program to promote the replacement of wood frame housing with steel framing. Steel is described as the environmental choice due to its high recycled con tent. "Red Pine Manufacturing, Co." does not have the resources to compete with this type of advertising program.*

2) *"International Wood, Inc." is establishing a factory 10 miles down the road in Lenora that will produce fixtures that will compete seriously with our moderately priced showcase line throughout the greater Twin Cities area.*

You should also include statements that identify the strengths and weaknesses of your company and product lines. These help the reader to understand exactly where your strategies are coming from.

Examples:

Strengths

1) *"Wood-N-Stuff" is recognized as producing the finest quality playground equipment in the nation.*

2) *Over the past three years, "Cabinets-R-Us" has managed to pay off all long and short term debt.*

3) *"Red Pine Manufacturing" has a close and long-term relationship with customers representing 85 percent of the market for red pine in the Great Lakes region.*

Weaknesses

1) *As our manufacture of wooden fixtures and millwork has become more technical, we are experiencing a critical shortage of employees with the training and experience to operate the new equipment.*

2) *Competitors are currently spending twice the amount of money on advertising in our key markets as WoodCo has budgeted.*

3) *Wood-N-Stuff chairs are not perceived as being higher quality even though they are priced higher Thus the price-conscious buyer is being lost.*

Finally, given this SWOT analysis, are there any significant issues that need to be addressed in the company's goals and objectives? If so, what are they?

Examples:

1) *Should Wood-N-Stuff stop making tables and instead buy them from outside sources for use in their dining sets?*

2) *Should WoodCo increase its advertising expenditures to match the competition?*

3) *Should Wood-N-Stuff undertake some marketing research to determine the markets for; and impact on, existing lines of a more moderately priced line of chairs?*

D. Goals & Objectives

At this stage you have summarized your situation as it now stands, and what's been going on recently. You've prioritized your strengths, weaknesses, opportunities, and threats, And you've raised some issues of some importance to the company. Now it's time to set your financial and marketing objectives.[14] These need to be very specific, and measurable!

Examples:

Financial

1) *Produce a net profit of $225 thousand in 1996.*

2) *Generate an average rate of return on investment (ROI) over the next 5 years of 15 percent.*

Marketing

1) *Produce sales of $5 million within 3 years.*

2) *Increase by 50 percent the number of customers who rank WoodCo's quality as being excellent.*

3) *Increase the frequency of contact with our customers to a minimum of one contact per week.*

4) *Raise the average realized selling price of all products by 3 percent.*

Remember, objectives should be realistic (but challenging), measurable, have some real or implied time period for accomplishment, be unambiguous, and should be stated in order of importance.

E. Marketing Strategies

Now that you've stated where you want to go, there are a number of ways you can get there. For example, to increase sales you can increase the average price per item, increase the number of items sold, sell more high priced items, or all of these.

Choose Best Path To Meet Objectives!

You're goal now is to choose the path that is most likely to achieve your objectives at the highest level of reward. This involves formulating strategies.

In forming strategies you are simply making decisions about choices, and defining those choices specifically. Effectively these will revolve around your marketing mix (4-P's), and your targeting and positioning strategy.

For Example:

<u>Target:</u> The price-conscious purchaser of chairs in the Twin Cities area, especially the female buyer.

Positioning: High quality/value relationship. Sturdy, yet attractive, for a modest cost.

29

Product: Establish new brand, with three styles of chairs in oak, maple, and pine. Simple form, sturdy construction.

Price: Price directly with competition.

Distribution: Primarily through furniture stores and major department store chains. Utilize existing sales force with new brand manager. Distribute to as many outlets as possible, and emphasize prompt delivery.

Promotion: Advertise both as point of sales and through brochures to the sales staff at the distribution outlets. Emphasize the quality/price relationship. Offer additional sales incentives to company sales staff and, in particular, to sales personnel at new outlets.

Finally, you want to establish some form of marketing research, either through your sales staff or some outside agency, to evaluate your performance and customers perceptions throughout the process.

F. Specific Action Steps

Now that you've described what is planned in a broad sense, it's time to get down to the details. In this section you want to describe exactly <u>what</u> you are going to do, <u>when</u> you are going to do it, <u>who</u> is going to do it, and <u>how</u> much the entire program is going to cost.

Examples:

1) *"Wood-N-Stuff" management will budget $5000 to hire an advertising agency to work with staff to develop a brochure for their new line of chairs. Bob Davies will be responsible for the development and completion of this project. A budget of $20 thou-*

sand will be set aside for the direct costs associated with the printing of the brochure, and this project will be completed by May 1st.

2) *A budget of $75 thousand will be set aside for costs associated with hiring, training, and the first year salary for a new brand manager for the new line of chairs. A commission based on total sales of this line will also be paid and is estimated at an additional $10 thousand, based on first year sales projections.*

3) *To meet the sales and profit objectives of the firm, the suggested retail price of the new line of chairs will be $99 per chain with a markup of 85 percent. Volume discounts of up to 20 percent off wholesale prices will be available to stocking distributors.*

G. Projections

This is one of the most straightforward sections. Using data gathered previously, you develop a budget to undertake the project as you deem best, and then develop an estimate of sales and profits that will result. It is especially important to note the process here. You DON'T estimate sales, and then estimate a budget to cover the costs. You do it the "other way around." Why?

Well, this is simply recognizing the fact that your sales are as dependent on what you spend to attain them as your costs are tied to the amount you sell. For example, increased spending in advertising assumes you will have increased sales. Estimating your budget first is the proactive approach, and recognizes your ability to control and impact your own results. You do the best you can, with what you have!!

Usually your projections are provided in a simple table, with a timeline, and this section would also include a profit and loss statement based on projected costs.

H. Feedback and Control

Now you need to discuss how you will monitor the process, and control your plan's implementation.

For Example:

- *Do you need customer surveys to measure attitudes?*

- *How will you gather and analyze sales data? Are existing methodologies sufficient?*

- *On what time schedule will you review your progress - Daily? Weekly? Monthly?*

- *What will constitute not achieving goals along the way? Being behind at all? Behind by 5 percent?*
- *What will trigger a major reconsideration of the project?*

- *What about contingency plans for things like strikes?*

- *Who makes the final decisions on any changes?*

These are just a few examples of the things you should consider. Remember, the more thorough your plan is up front, the more likely the project is to be successful. Don't kill a year's sales by skipping an hour of thoughtful brainstorming about what might go wrong, or what you might do if things go better than expected. The key is:

Be Prepared!

NoteBook

The KEYS To Success!

Part IV. BUSINESS PLANS

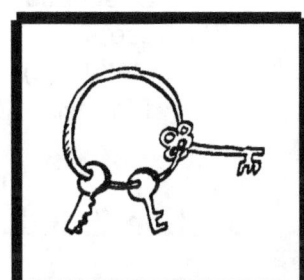

As mentioned earlier, the primary difference between Business Plans[15] and strategic or individual marketing plans is that the business plan is generally designed to be read by someone outside the day-to-day operation of the company (for example: your investors or your bank). For this reason it requires additional detail about the firm that is unnecessary in the previous two plans. In addition, the business plan should include a marketing plan when a new activity (e.g. starting new firm or expanding existing. one) is involved.

A complete example of a business plan is provided in Appendix B. This example is designed for a particular industry, but careful review should provide ideas on how a firm could fit its own detail into the format. In this section we will give a few comments to supplement the example provided.

In general a business plan includes:

1. Executive Summary
2. Introduction
3. Industry Overview
4. Description of your Company
5. Marketing Plan
6. Organizational Plan of Company
7. Operating Plan
8. Financial Plan
9. Limitations

Banks often Require Business Plan!

Section 1. Executive Summary

This is a crucial part of your report! It should be written **last,** and should summarize the "heart" of what the entire report is about. Enough detail should be given that the reader's interest is aroused, but it also should be concise enough that it can be read in 5-10 minutes. The major components that should be included in this summary are:

- Company name and location (include any significant benefits of the location!)
- Descriptions of the industry (brief)
- Purpose of the business plan
- Description of the product (service)
- Description of the market (include projections)
- Financial projections
- Finance amount sought
- Payback period
- Goals and objectives of the company

- Marketing strategies to achieve goals
- Company expertise (management's as well)
- Impacts on community and/or region

Throughout this summary you should emphasize your successful characteristics wherever possible.
For Example:

Wood-N-Stuff is the largest manufacturer of wooden playground equipment in the state of Minnesota.

Cabinets-R-Us boasts the largest selection of Cherry cabinets in the United States.

Last year Michigan Superior Distribution had the highest profit margins in its 80 year history.

Section 2. Introduction:

In this section you are simply introducing the reader to your company, your industry, and your needs. In some ways this is simply to give the reader an idea of what to expect in the balance of the report. This section should also include at least the following three areas:

A. Purpose of the Business, including:

- General description of the company
- General description of the product/service
- Company goals

B. Broad Overview of the Industry, including:

- General description of the industry
- General description of the market
- Brief discussion of the competition

C. Financial Summary, including:

- Description of funding sought and its use
- Method(s) of financing
- Financial goals

Section 3. Industry Overview

You should provide your readers with a detailed description of your industry. Remember, this plan is aimed for those outside your company, so the better these people are informed, the more likely they are to support your endeavors. Utilize university extension faculty, industry association staff, and local libraries to supplement your companies direct knowledge of your industry. You would be surprised at how much detail is out there if you only look. This is extremely important if you are applying for financing. The better you do your homework, the more likely financiers will be impressed. Remember to include:

A. The history of your industry, how it developed, its current size and makeup, factors that have led to its growth, and a description of how your company developed in this context.

B. Trends within your industry, areas of growth, and current industry-wide research and development (R&D) efforts.

C. Key success factors, including answers to the following:

- How do you differentiate your company from the rest of the industry?
- What distinctive features do your products/services have?
- What are your competitive advantages?

- Do you conduct any R&D activities, and if so, in what areas?

D. Your company's segmentation, targeting, and positioning strategies (see Part III, Stage 1, for discussion of these activities).

Section 4. Product, Service, and Process

In this section you should provide exact details of the product, processes and services you are planning. In some cases banks may ask outside consultants to review your plans to be sure they are realistic. The opposite may be true as well - it may not be an expert reviewing your plans, so leave out the jargon and be concise. Be sure to include drawings, graphics, and pictures to facilitate understanding. This is especially important if you are expanding internationally. Your plan should have:

A. Detailed description of the product, including:

- Important customer benefits,
- Intended quality level,
- Intended use,
- Intended price/performance relationship,
- How it fits into your current product "portfolio,"
- Technical developments involved,
- Regulatory status, e.g. EPA requirements, if applicable, and current status, and
- Any other considerations that influence the plan.

B. Description of the services you provide, including:

- Warranties,
- Marketing support,
- Technical support, and
- Any other "field" support you provide.

C. Detailed description of the processes involved, including:

- Flow diagram if applicable,
- Major machinery and equipment, and
- Any correlation between equipment size/type and volume/quality.

Section 5. Marketing Plan

A marketing plan should be included *only for new product(s) and/or new market(s) related directly to the project at hand.* For example, if you are seeking financing for a new CNC router, then provide a plan only for the new products you will be able to add or new markets you plan to serve as a direct result of the new machinery. You generally do not need a marketing plan in situations in which you are simply replacing existing equipment. However, if a plan is required, it can be developed using the same process described in Part III.

Section 6. Organization of your Business

In this section you describe the structure of your company, and you should include any anticipated changes (such as additional management personnel) as your company grows. Basically this section includes:

A. Formal structure of your company, including:

- Tax status, and
- Legal attributes

B. Management team and staff

- Use organizational chart.
- List key personnel (including Board of Directors!), and include their resumes in appendix.
- List any business advisors you use regularly and their specific skills.

C. Ownership of company

- List principal equity owners and their affiliation(s).
- Describe how owners are involved in management.
- Describe any public funding.

Section 7. Operating Plan

The operating plan simply describes how you plan to manage your company in both the short and long term. This includes management of the plant, physical equipment, and your employees. It should be a realistic description of your activities, including:

A. Description of the company's location(s), including:

- Any special abilities or requirements,
- Why you chose the location(s), and
- Facility costs - including a breakdown of per foot costs.

B. Description of the labor force, including:

- List of names of management,
- Salaries,

- Description of labor positions, annual wages, and a comparison with prevailing wages in the local area,
- Description of fringe benefits and costs,
- Description of workers' union(s) if applicable,
- Discussion of any staff additions,
- Discussion of stability of positions (seasonal, cyclical, etc.), and
- Description of any special skills needed, or training provided.

C. Describe how you will access raw materials, including:

- Sources,
- Projected costs,
- Freight methods and costs,
- Description of factors that may potentially influence availability of these materials, and
- Description of anticipated terms of purchase, lead times, and industry standards for these items.

D. Provide an anticipated production schedule, and also include:

- Description of expected delays to start-up,
- Description of how you will account for "work in process," and
- Description of your inventory control procedures.

E. Describe long-range plans, including:

- The ultimate goal of the business,
- Plans to achieve that goal,
- Your planned business environment, and the work attitudes you desire to foster, and
- A discussion of possible expansions, diversification, and additions.

Section 8. Financial Plan

Development of a financial plan is extremely difficult without either an accounting background, or the help of an accountant. Basically, you need to develop a detailed 5 year plan, and you need to provide a source to justify your estimates (be sure to use reliable sources!). Generally there are 5 parts to a financial plan (remember the example in Appendix B). These are:

A. A listing of the capital requirements, sources of information, contingencies, and reserves.

B. A description of your financing plan, including all major alternatives considered and all sought. Describe all sources of capital.

C. A beginning balance sheet (current if presently in business, pro forma for a new business).

D. A complete statement of projected operations and cashflows. You should:

- Include monthly data for year 1, quarterly for years 2 and 3, and annually for years 4 and 5,
- Separate plan into sales and financial sections,
- Explain assumptions in footnotes,
- Discuss how costs may fluctuate with production volumes, and
- Describe the cost system and budgets you will use.

Note: A list of the formulas used by Pepke (1993) is included in Appendix C. The formulas correspond to the headings used in the sample pro forma in Appendix B.

E. Provide a discussion of the investment criteria that you use, including calculations for:

- Internal rate of return,
- Break-even point,
- Present net worth,
- Ratio of present net worth to initial investment,
- Any other ratios requested specifically by your audience, and
- Sensitivity analysis, showing changes in interest rates and their impact on your figures.

Section 9. Limitations

In this section you want to show the readers that you are realistic in your expectations, and that you are trying to anticipate what might go "wrong." In this section you should discuss the impacts of:

- Changes in economic climate,
- Adverse market trends, and
- Increased Costs.

In addition, you should discuss the accuracy of your assumptions, and project your company out 10 years to describe your long-term expectations.

NoteBook

The KEYS To Success!

Part V. SUMMARY

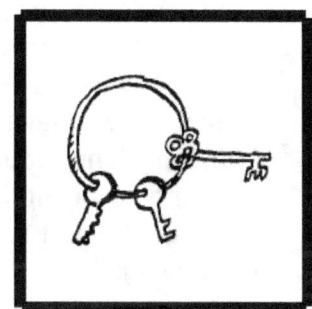

As noted earlier, the purpose of this Guide was to assist you, the wood products manufacturing executive, in making your business life a little easier, improving your chances of success, and maximizing your firms capabilities. It is worth repeating that research has shown that a formal planning process is a key element in the success of wood products manufacturing companies, especially in regard to development of new products and new markets. The existence of formal, written plans allows firms to create strategies and programs to implement their ideas, and facilitates their ability to measure and control their results.

It should be remembered that **Strategic Planning** is engaged in **creating the future,** rather than predicting the future. The support of top management is critical to the success of the planning effort. Although most innovative managers of wood products firms tend to think and act strategically, not all are particularly successful. Formal strategic planning increases the manager's chance of success.

The Marketing Plan is the blueprint for the construction of all marketing programs. It deals primarily with strategy as it relates to the company's marketing mix, often referred to as the 4-P's of marketing: Product attributes, Pricing strategies, Promotion

Formal Plans Are Important!

strategies, and Place (distribution methods). It should be remembered that marketing planning concentrates narrowly on a firm's specific product and/or market.

Both the **Marketing Plan** and the **Strategic Plan** are designed primarily for internal company use, although stakeholders outside the firm can and should be involved at different stages of the creation of such plans. On the other hand, a **Business Plan** should be developed specifically with the stakeholder in mind, and usually for those outside the firm, such as bankers, potential partners, investors and others your company is trying to influence. Keep in mind that a business plan is not only a comprehensive gathering of detailed information on your business (or potential business), but also a way of **monitoring** the business.

This Guide, if used properly, should be viewed as a first step in your planning efforts. Additional information on the three "plans" is available from textbooks, business magazines, workshops and seminars, and trade organizations. The references used in writing this Guide are excellent sources of additional information for individuals wanting more detailed assistance in their planning efforts.

Remember, your plans should not be set in stone but are, to some extent, a work in process. This doesn't mean you are continually changing your goals and objectives.

However, you should be monitoring your progress toward your goals and adjusting your strategies and programs to assure that you meet or exceed your objectives.

Lastly, as we have mentioned before, the plan (or plans) you develop are dynamic, and should be updated periodically to be of the most benefit to your firm. Consider the old saying:

Those who fail to plan....plan to fail.

Good Luck!!

EndNotes

[1] Howe, Gephart, and Adams, 1995.

[2] Much of Part II was adapted from Bengston, 1991.

[3] This section was adapted from Goodstein et al., 1993.

[4] The mission statements in this section were adapted from Buechler, 1993.

[5] Kotler, 1991, p. 4.

[6] Pride and Ferrell, 1991, p. 5.

[7] Levitt, 1960, pp. 45-56.

[8] Myopia here refers to "short sightedness."

[9] Zimmerman, 1955, p.48.

[10] Kotler, 1991, p. 62.

[11] Kotler, 1991, p. 63.

[12] Cooper, 1991, pp. 183-191.

[13] Contribution margin is the portion of sales each unit contributes toward what is generally considered as fixed costs.

[14] Remember, goals can be general, but objectives are specific!

[15] Much of Part IV was adapted from Pepke, 1993.

References and Recommended Reading List

Bengston, D. N. 1991. Strategic planning in public forestry research organizations. In Research management for the future, tech. coor. D. P. Burns, 107-108. General Technical Report NE-157. Radnor, PA: USDA Forest Service, Northeastern Forest Experiment Station. [Article]

Buechler, Oris C. 1993. Business plan development and analysis handbook. Northeast Utilization & Marketing Council and USDA Forest Service, State & Private Forestry, Radnor, PA. [Very detailed coverage of the subject; excellent source of economic, industry, and market information data.]

Cesa, Ed. 1992. A marketing guide for manufacturers & entrepreneurs of secondary-processed wood products in the Northeastern United States. NA-TP-09-92. USDA Forest Service, State & Private Forestry, Morgantown, WV. [Easy-to-read with many color photographs; "locating customers" is a major focus of handbook; names and addresses of "additional sources of information" are provided for manufacturers and entrepreneurs in the 20 northeastern states.]

Cooper, Robert J. 1991. Product innovation in the timber industry: The challenge for the 1990's. Journal of Wood Science, 12(3): 183-191. [Article; Cooper has done extensive research into successful new product introduction.]

Goodstein, Leonard D., Timothy M. Nolan, and J. William Pfeiffer. 1993. Applied strategic planning: a comprehensive guide. McGraw-Hill, Inc.: New York. [Thorough analysis of strategic planning; includes detailed description of "values".]

Howe, Jeffrey L., J. Gephart, and R. Adams. 1995. Identifying the success characteristics of small and medium size manufacturers: A survey of wood products companies in Minnesota. Journal of Applied Technology, St. Thomas College, St. Paul, MN, Fall. [Article]

Kotler, Philip. 1991. Marketing management: Analysis, planning, implementation and control. Prentice-Hall, Inc.: Englewood Cliffs, NJ. [Kotler is considered by many to be the leading authority on marketing; textbook is very comprehensive.]

Levitt, Ted. 1960. Marketing myopia. Harvard Business Review, July-August issue.

[Article; Levitt is one of the "god fathers" of marketing.]

Mater, Jean, M. Scott Mater, and Catherine M. Mater. 1992. Marketing forest products. Miller Freeman, Inc.: San Francisco.
[Hands-on book that offers ideas for increasing market share and capturing market segments; written for marketing newcomers as well as experienced strategists.]

Pepke, Ed. 1993. How to write business plans for forest products companies. NA-TP-17 USDA Forest Service, State & Private Forestry, St. Paul, MN.
[Step-by-step business plan guide; examples, sources of assistance, and information deal specifically with the forest products field.]

Pride, W.M. and O.C. Ferrell. 1991. Marketing concepts and strategies. Houghton Mifflin: Boston.
[Introductory textbook; describes all marketing activities in some detail.]

Sinclair, Steven A. 1992. Forest products marketing. McGraw-Hill, Inc.: New York.
[Comprehensive textbook that includes examples of real companies as well as real data on major markets; includes sections on "international marketing" and the "nature of competition".]

U.S. Department of Commerce. Statistical Abstract of the United States 1994: The National Data Book. Bureau of the Census, Washington, DC.
[Statistical data compiled from various national and international sources.]

U.S. Department of Commerce. County Business Patterns 1992 (by state). Bureau of the Census, Washington, DC.
[Provides annual subnational data by two-, three-, and four-digit levels of the Standard Industrial Classification (SIC) system; useful for making basic economic studies of small areas and for analyzing the industrial structure of regions.]

U.S. Department of Commerce. (In press). 1992 Census of Manufactures. Bureau of the Census, Washington, DC.
[For each industry (industry series), data on value of shipments, value added by manufacture, capital expenditures, employment, and payroll are shown by employment-size class of establishment, state, and degree of primary product specialization. For each area (geographic area series), data presented for industry groups and industries on value of shipments, cost of materials, value added by manufacture, employment, payroll, hours worked, new capital expenditures, and number of manufacturing establishments for the state, counties, and selected places.]

43

Williston, Ed M. 1991. Value-added wood products: Manufacturing and marketing strategies. Miller Freeman, Inc.: San Francisco.
[Hands-on guide for sawmills, plywood, panel, and other forest product plants; explains how to uncover opportunities for increased return in every phase of the business.]

Zimmerman, M.M. 1955. The supermarket: A revolution in distribution. McGraw-Hill: New York.
[Article]

Appendices

Appendix A

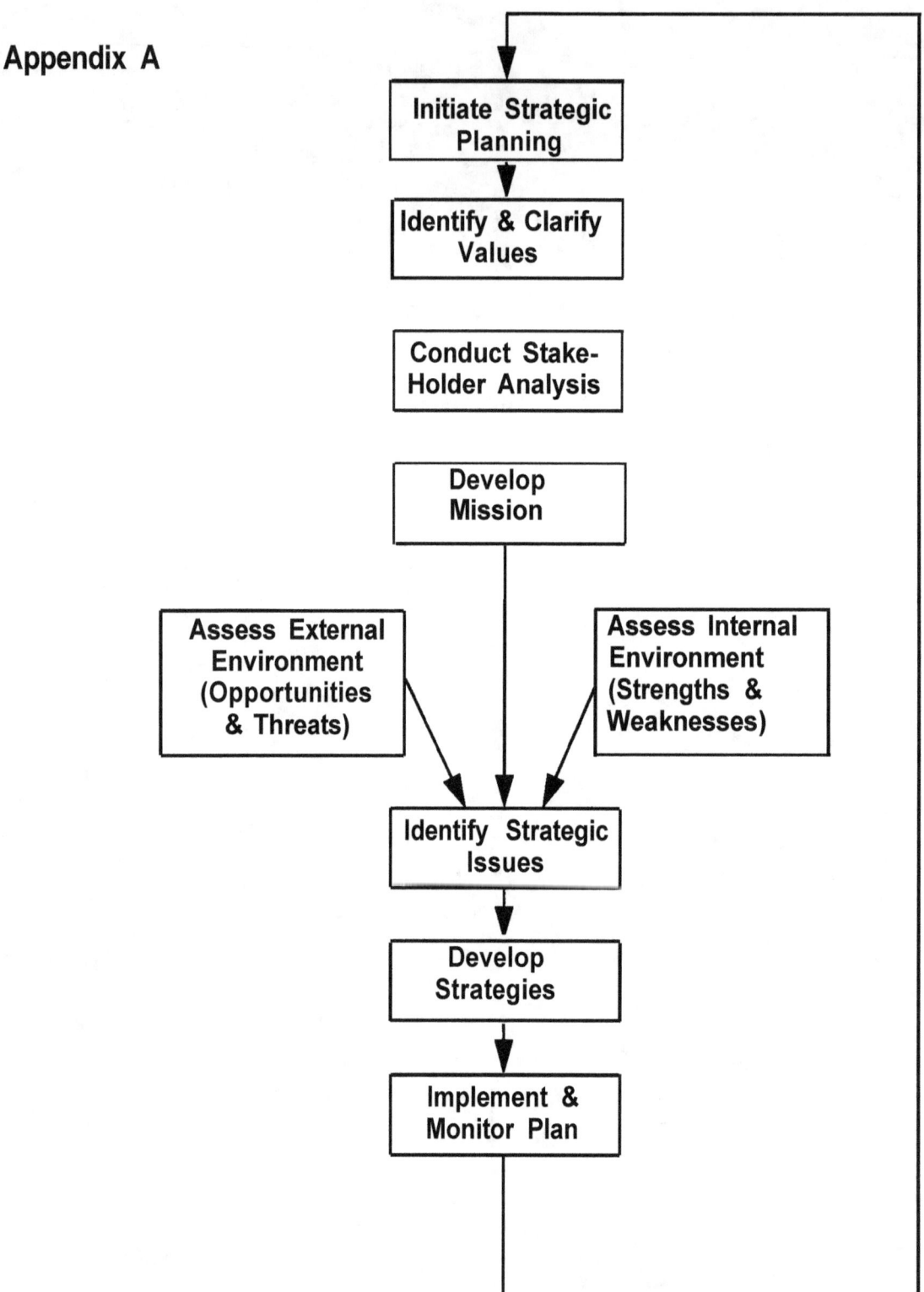

A Strategic Planning Process
Source: Adapted from Bengston (1991)

Appendix B

READY-TO-ASSEMBLE FURNITURE MANUFACTURING
A BUSINESS PLAN
FOR THE NORTHEASTERN AREA

Prepared by:

Ed Pepke
Marketing Specialist
Forest Management and Utilization
Northeastern Area, State and Private Forestry
USDA Forest Service
1992 Folwell Avenue
St Paul, Minnesota U.S.A.

August 1988

NA-TP-12

CONTENTS

READY-TO-ASSEMBLE FURNITURE MANUFACTURING
A BUSINESS PLAN
FOR THE NORTHEASTERN AREA

EXECUTIVE SUMMARY

Ready-to-assemble (RTA) furniture is the fastest growing segment of the world's furniture market. RTA furniture is shipped unassembled to eliminate assembly costs, to facilitate transporting, and to reduce shipping costs-the savings and the assembly are passed to the consumer. The unassembled furniture is packaged in boxes to minimize freight damage. Previously known as knock-down (KD) furniture, RTA furniture has become the consumer friendly furniture of the 1980s because of material, hardware, and design innovations. Industry estimates the 1987 U.S. retail market for RTA furniture was $2 billion retail, and *Wood & Wood Products* predicts that figure will double by 1990.

RTA furniture typically is constructed from particleboard coated with colored melamine or wood veneers. Entertainment centers, electronic and computer furniture, and storage units are the most popular pieces of furniture purchased. At one end of the market is the intensely competitive low-priced RTA furniture with its good sales record, and at the other end is high-priced RTA furniture with its limited affordability and smaller, more exclusive market. This business plan segregates a market niche for medium-quality, moderately priced RTA furniture. While the plan is based on production of a single item of furniture, a simple bookcase, it is assumed that a manufacturer, such as the hypothetical RTA, Inc., of this report, would expand into compatible pieces of furniture. The flexibility of RTA furniture enables a simple bookcase to be transformed into a chest of drawers, a closed cabinet, an entertainment center, or a desk by adding the appropriate drawers, doors, shelves, and hardware.

The costs in the "Statement of Projected Operations and Cash Flows for the First 12 Months" are the quoted prices for plant, machinery, equipment, raw material, and supplies. The plan reflects a conservative approach: when estimates varied, the higher costs were used. The plan is pro forma; interested investors should regard it as a starting point and adapt these ideas and adjust the costs for their individual uses.

The values in the financial statements for the first five years of operation were obtained through the EVALUE computer program. EVALUE is a microcomputer program for determining the financial feasibility of a forest products industry investment, which was developed at the USDA Forest Service's Forest Products Laboratory in Madison, Wisconsin. Based on projected incomes and expenses, the program makes the financial analysis in light of the applicable tax laws and yields the decision criteria such as present net worth, payback period, and internal rate of return. Using relatively conservative input values, EVALUE predicted an internal rate of return of 33.7 percent and a payback period of 3.2 years for the hypothetical company.

The business plan alludes to financial incentives available to new or expanding companies through the 20 state and community governments of the USDA Forest Service's Northeastern Area. Frequently these incentives are land grants, building grants, tax allowances, financial grants, and low-interest loans. A list of each state's offerings is included in Appendix A.

An entrepreneur must customize the ideas, numbers, and methods in this pro forma business plan for each new set of circumstances encountered in establishing an RTA furniture manufacturing facility. The raw materials, machinery, and labor are available in the Northeastern Area of the United States; the current demand for RTA furniture is strong and growing. The product has flexible designs that allow multiple pieces to be made from the same basic case configuration. A small-scale production level is proposed in this plan for a minimal investment, but it easily could be geared up to a much larger production operation.

Parts of this business plan may be too simplistic for an established manufacturer who wishes to expand into the Northeastern Area. For that manufacturer, the sections on the market size and trends are still of value.

INTRODUCTION

The Northeastern Area is a 20-state area designated by the United States Department of Agriculture's (USDA) Forest Service (see inside front cover). Over one-half the population of the United States resides in the Northeastern Area, and other major markets border the Northeastern Area. This pro forma business plan for RTA, Inc., is applicable to a new manufacturer of ready-to-assemble (RTA) furniture, and this report's flexibility allows the manufacturer to customize the plan. Portions of this plan, such as the market analysis, are of interest to existing furniture manufacturers.

RTA furniture is the fastest growing segment of the world's furniture market. RTA furniture has evolved from lower quality knock-down (KD) or quick-assembly furniture to what is known as consumer friendly furniture of the 1980s. Today, through material and technological innovations, RTA furniture manufacturers have improved their products while restricting prices to a fraction of those charged by the manufacturer of fully assembled furniture. RTA furniture is shipped unassembled to minimize shipping costs and total furniture costs; the savings and the assembly are passed on to the consumer. Shipping losses are less with RTA furniture because the flat, packaged furniture is much less vulnerable to damage.

A. Purpose of the Business

This business plan supposes an RTA furniture manufacturer who wants to produce a line of medium-quality furniture that would be marketed through mass merchandisers, specialty RTA furniture stores, and home building centers (large building supply stores that sell some furniture to the do-it-yourself (DIY) market). In the initial production stages, RTA, Inc., purchases veneered panels, machines these panels to designed dimensions, finishes, and packages and ships the unassembled units. Business expansion could take a variety of routes, depending on the management and market. A northeastern U.S. firm competes primarily with imported RTA furniture from the Pacific Rim countries and Europe. This domestic firm will have the advantage of being closer to growing major markets.

B. Industry Overview

In the United States, retail sales of RTA furniture are estimated to be $2 billion per year and are increasing. Most furniture imported into the United States is RTA, and most RTA furniture manufacturers are outside the United States. The domestic U.S. RTA furniture manufacturing industry has a few quality manufacturers trying to keep up with the demand for RTA furniture. Forty percent of Great Britain's total 1985 furniture sales was in RTA, while only 3 to 5 percent of the comparable U.S. market was in RTA. The U.S. RTA furniture market has been growing in excess of 5 percent annually since 1977. In the November 1987 *Wood & Wood Products*, the authors estimated a doubling of the U.S. retail RTA furniture market by 1990.

C. Financial Summary

This plan for RTA, Inc.'s furniture plant requires an investment of $1.5 million to establish a small manufacturing operation. The initial investment easily could be scaled up to meet the larger production plans of some investors. In addition to some funding from the investors, the industrial development incentives from local, state, and federal governments could be substantial in terms of land, buildings, and capital.

A. History

Originally RTA furniture was known as KD for knock-down and unfortunately the products came apart almost as easily as they were put together. Substantial improvements in construction methods and hardware have heightened product quality to its present consumer friendly status of the 1980s. While RTA furniture has a long history in Europe, the early introductions into the U.S. market were poorly constructed compared to today's products. The original products were vinyl-wrapped particleboard television stands and bookcases which sold for under $40. The products *were* purchased as short-term, disposable furniture, but the value was still good for the price paid. The industry realized that long-term survival would entail increasing quality, design, and function. Improvements have enabled composite board to be printed with wood grains or colored with coatings of paint or melamine. Construction methods, hardware, and fasteners have improved dramatically in the last decade.

RTA furniture (also called flat-pack or carry-home furniture or furniture-in-a-box) is made of solid wood, wood veneers, laminated wood cores, particleboard, medium density fiberboard (mdf), low-pressure paper laminates, high-pressure laminates, melamine, vinyl wrap, or metal. Products include wall systems, bedroom furniture, living room furniture, cabinetry, and other household and office furniture. Current technology requires use of the 32 millimeter (mm) system, which uses construction and hardware on 32mm spacing. The two benefits of the 32mm system design standardization are (1) a reduction in boring machine set-up time and (2) the versatility of the basic cabinet box that can be altered by using interchangeable parts. Since assembly is not required and the machinery is automated, RTA furniture is produced with less labor than conventional furniture. The initial investment in equipment is recovered through reduced production and labor costs. Packaging the product in boxes is the most labor intensive step.

The primary distribution paths for RTA furniture are the mass merchandisers such as K-Mart, Sears, J. C. Penney, and Montgomery Wards and discount stores like Target, Caldor, and Gemco. Well-known retail chains, who often carry two or more lines of RTA furniture to segregate themselves from the mass merchandisers, are Storehouse, Workbench, Room & Board, and Conrans. Other RTA furniture outlets are contemporary specialty chains, including small lifestyle shops and Scandinavian stores importing RTA products direct from Europe. One of the newest markets for RTA furniture is home improvement stores; these stores often carry RTA wall units and cabinets, bathroom vanities, and kitchen cabinets.

B. Trends and Projected Growth

Retail sales in the United States are estimated at $2 billion per year and are increasing. According to the Ready-to-Assemble Furniture Association, 3 to 5 percent of all furniture sales are consumer assembled. Forty percent of Great Britain's furniture sales are in RTA–there exists tremendous potential for the U.S. market. U.S. retail furniture sales for 1987 were about $19.2 billion, which is about 5.3 percent greater than 1986; imported furniture, most of which is RTA, represented 18 percent of this total in 1987. Glen Goodwin, a Partner in Seidman & Seidman, the major accounting firm for the furniture industry, predicted no increase in sales in 1988 for the domestic furniture manufacturing industry and a 14 percent increase (down from 1987's increase) in imported furniture sales in 1988. Additional information on market trends appears in Part 3, Section A, "Market Size and Trends".

An article in the May 1986 *Wood & Wood Products* states that the DIY market is now the fastest growing segment in the retail economy. For 1986 an 8.6 percent growth rate is projected for DIY versus a 6.6 percent growth rate for overall retailing. Kitchen and bath remodeling are the largest and most profitable projects in the home improvement industry, with a conservative estimate of $15 billion in annual sales. Francis Jones, Executive Director of the National Kitchen & Bath Association, (NKBA) states in the February 1988 *Wood & Wood Products,* "The interest deduction for home equity loans, increasing income, decreasing unemployment, the continued aging of 100 million housing units and the decrease of new housing units are all positive reasons for increased kitchen and bath sales." NKBA sees an 11 percent increase in kitchen remodeling and a 14 percent increase in bath remodeling in 1988. The National Kitchen Cabinet Association, in the March 1988 *Wood & Wood Products,* predicted 43.1 million cabinets for kitchens and baths would be sold in 1988 (a modest 0.7 increase over 1987). Housing starts, 1.6 million units in 1987, are estimated by the National Forest Products Association to be 1.5 million units in 1988. *Wood & Wood Products* feels that the European influence, in terms of design and manufacture, will result in many of these cabinets being manufactured under the 32mm system and in RTA form.

C. Key Success Factors

The strong current market for RTA furniture, resulting from the low interest rates that spur consumer spending, home building, and new household formations, has increased imports of RTA furniture to the United States; also contributing to increased imports is the currently devalued U.S. dollar.

Much imported RTA furniture is relatively low quality, while the remainder of the imports and the small domestic production are higher quality with an accordingly higher price. The market is open for a medium-quality product that appeals to the quality- and value-conscious buyer who does not find the highest priced RTA furniture affordable. Production of a medium-quality product allows RTA, Inc., to expand either up to a high-end product or down to a low-end product, depending on future market conditions.

New machinery tooled exclusively for the 32mm method of construction is available. The advantage of the 32mm system is the standardization of spacing which enables easier set-ups and quicker assembly. Hardware developed for the 32mm system has advanced sufficiently so that fasteners, hinges, handles, drawer glides, and other pieces of hardware match the 32mm spacing. Versatility is imperative in manufacturing RTA furniture, and the 32mm system machinery provides the ultimate flexibility.

D. RTA, Inc's Fit in the Industry

Furniture buyers at Room & Board and International Design Center in Minneapolis and St. Paul, Minnesota, agreed that there are two segments in the retail RTA furniture market. The low-price market (or KD lifestyle) is shopping for price only; the anticipated usage is for 2 to 3 years and often for children. The high-quality market is shopping for quality, durability, function, and appearance. More wood is visible for the high-quality market. Price is less important, and frequently the customer has the store assemble and deliver the furniture.

Del Olson, owner and RTA furniture buyer for International Design Center, feels that he must carry at least two models of every piece of furniture in order to sell anything. Olson says, "I can't sell a black table unless the customer has two black tables to choose from." Similarly, he carries one line of low-price RTA furniture for comparison with his lines of higher quality RTA furniture. At International Design Center, the hang tags read "Furniture-in-a-Carton." The customer can take the furniture home in a box or pay for assembly and delivery. Room & Board will not sell their high-quality Techline, manufactured by Marshall Erdman near Madison, Wisconsin, unassembled-they like to keep control by in-house assembly and gain the additional revenue. However, Room & Board does sell their lower line Star, which is from Europe, assembled or unassembled. Discount merchandisers and department stores commonly sell all RTA furniture unassembled. Many furniture manufacturers from the Pacific Him countries ship their furniture to captive assembly plants in the United States. Most imported furniture is shipped unassembled to some degree.

If a firm contemplating following this business plan has previous experience in the high-end RTA furniture market, it should expand in that direction. Competition in the low-priced RTA furniture market is very tough, and any firm would always compete with the high-production manufacturers that export to the United States.

The fledgling firm that has little or no experience in the RTA furniture market should produce a medium-quality product priced to fall between the two extremes. This product would appeal to specialty stores' buyers who are trying to find a product above low-end, cheap RTA furniture and below top-end, expensive RTA furniture.

A. Product Description

RTA furniture typically is constructed from particleboard with colored melamine or wood veneer surfaces. Often the products produced have square corners, as the most popular products are casegoods. The furniture is shipped unassembled in a box and the final assembly is performed by the retail store or customer.

As previously described, the RTA furniture produced by RTA, Inc., is of medium quality, below top-of-the-line RTA furniture (with its associated top-of-the-line prices) and above lower quality, cheap RTA furniture. In contrast with low-priced RTA furniture, the medium-quality product has thicker board, heavier hardware, and better machining, finishing, and edgebanding. Drawer and door hardware has evolved due to the 32mm standardization and the hidden camlock system of assembly (a quarter turn of the cam locks pieces together which eliminates turning screws). The firm should produce a line of furniture with components to fit the bedroom, living room, kitchen, and office. The product in greatest demand is the wall system of storage shelves, which adapts to any area of the home or office. Shelf storage units can be fitted with a variety of accessories: solid or glass doors to produce cupboards, drawers and fold-down tops to make desks, and drawers to make chests. RTA, Inc., designs products with as much versatility as the machinery, materials, and hardware permits and with as much variety as the market demands.

B. Service Description

The RTA furniture manufacturer ships the product unassembled, in a box to mass merchandisers, specialty stores, and home improvement centers. The manufacturer provides detailed instructions for assembly to the final customer as well as the store. The manufacturer also provides some sales support to store managers by educating them about the manufacturing process, construction materials, assembly, and product quality. The manufacturer publishes a catalog showing the various furniture pieces with illustrations depicting the versatility of usage. The catalog with other promotional materials helps market the products through the retailers. The manufacturer and retailers must agree about some type of repair or replacement service for parts damaged in shipment or use.

C. Process Description

This process is for a new company establishing the minimal capital system to produce medium-quality RTA furniture (fig.1). The plan could be scaled up considerably for firms

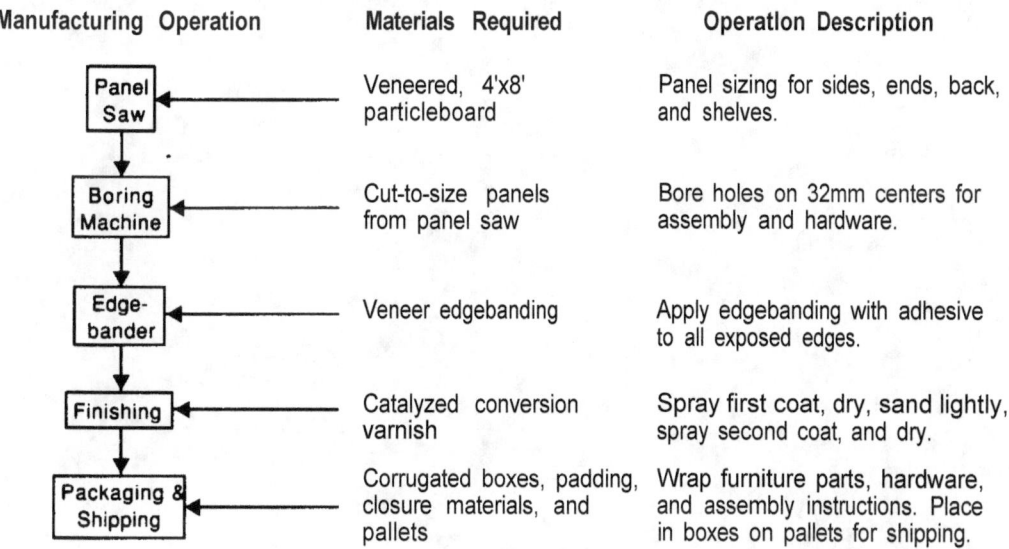

Manufacturing Operation	Materials Required	Operation Description
Panel Saw	Veneered, 4'x8' particleboard	Panel sizing for sides, ends, back, and shelves.
Boring Machine	Cut-to-size panels from panel saw	Bore holes on 32mm centers for assembly and hardware.
Edge-bander	Veneer edgebanding	Apply edgebanding with adhesive to all exposed edges.
Finishing	Catalyzed conversion varnish	Spray first coat, dry, sand lightly, spray second coat, and dry.
Packaging & Shipping	Corrugated boxes, padding, closure materials, and pallets	Wrap furniture parts, hardware, and assembly instructions. Place in boxes on pallets for shipping.

Figure 1 .—*Process chart for RTA furniture.*

wishing to process more than the 100 sheets of particleboard per shift that is described here.

Several equipment manufacturers were solicited for general quotes on their machinery: because SLi Machinery Corporation of Grand Rapids, Michigan, was the only company to respond, its machinery and costs will be used. This use does not constitute endorsement of any particular equipment, and users of this report are encouraged to solicit bids from a number of manufacturers of 32mm equipment.

The first piece of machinery that comes to mind when thinking about RTA furniture manufacturing is the panel saw. This machine is a sophisticated table saw used to accurately size cuttings from large sheets of panel products, in our case veneered 4-foot by 8-foot particleboard. A new operation needs a versatile machine capable of panel sizing, tenoning, and dadoing. SLi's TS 100 has the required flexibility and capabilities and has a additional scoring saw preceding the main saw.

While not used during the first step in RTA furniture manufacturing, the boring machine is the heart of the 32mm system. SLi's Biesse Forecon 51 is a compact multi-spindle boring machine that performs both horizontal and vertical boring without changing position of the boring heads. The horizontal unit has 21 spindles, and the swiveling vertical units have 15, 18, or 21 quick-disconnect boring heads. This small, versatile unit could be used eventually for off-line boring if the company expands and acquires larger equipment for on-line production.

The edges of composite board products require edgebanding of wood veneer, Formica, PVC (plastic), or wood edgings up to three-fourths of an inch in thickness. SLi's Nova 2 is a single-sided edgebander with a hot melt glue pot, an end cutting unit, and top and bottom trimming heads. The

trimming head angles are adjustable to 35 degrees if beveling is desired.

Finishing the cut-to-size, bored, and edgebanded panels necessitates a cost-effective finish system. For RTA, Inc., a relatively low-cost plant, compatible finish equipment includes an airless sprayer with a compressor and a pot for the finish. An airless sprayer operates at 5 cubic feet per minute which is one-half to two-thirds less air pressure than conventional air guns; the advantages are finish material savings from reduced overspray, thicker films that reduce the number of coats required, energy savings, and less filter maintenance of the booth. A spray booth with an explosion-proof fan motor and a filtered air exchange system should meet most fire codes. The flat panels can be sprayed on racks, which also facilitates air drying. The finish itself is a catalyzed conversion varnish that provides a very adequate finish in two coats (with an intermediate step involving sanding). To eliminate any final rubbing or polishing, the finish itself can provide the desired final sheen.

The finishing system described above is relatively labor-intensive compared to automated systems for flat-line finishing that are available now. With an increase in production volume, a more automated system might be cost justified in the future. Automated systems involve automatic spray, roll, or curtain coating, force dry facilities, buffers or sanders, and conveyors. New finishing technology such as electrostatic spraying should be investigated if production is being significantly upscaled from this plan.

Since no assembly is required, the fully finished parts are packaged in corrugated cardboard boxes and shipped on pallets. If furniture is ever sold assembled or partially assembled, the manufacturer needs a doweling machine and case clamps.

A. Market Size and Trends

RTA furniture is the fastest growing segment of the world's furniture market. Retail U.S. sales volume is increasing from $2 billion per year. Only 3 to 5 percent of U.S. furniture is consumer assembled, while 40 percent of Great Britain's furniture sales are in RTA furniture. 1987 sales of all furniture totaled $19.2 billion on the retail level; no increase in domestic manufacturers' sales is predicted for 1988 while a 14 percent increase is projected for 1988 for imported furniture retail sales. A study by the Minnesota Department of Trade and Economic Development cites over 5 percent growth annually in the RTA furniture market since 1977. The State of Minnesota study projects an annual growth rate of 4 percent or more to the year 2000. "If we were to have a year where we only had a 25 percent growth in sales, I'ld (sic) tell you we were having a bad year," stated Tim Wilson, Vice President of Affordable, one of the faster-growing RTA furniture manufacturers interviewed by *Furniture/Today* Newspaper. Wilson said sales growth from 1985 to 1986 ranged from 20 to 70 percent for RTA furniture companies. Positive factors leading to growth are low current market saturation, highly competitive pricing, innovative production technology, and maturing distribution channels. The dominant industry trend is towards improved design and increased quality. *Wood & Wood Products* magazine in November 1987 predicted a doubling of the $2 billion retail sales by 1990 (fig. 2).

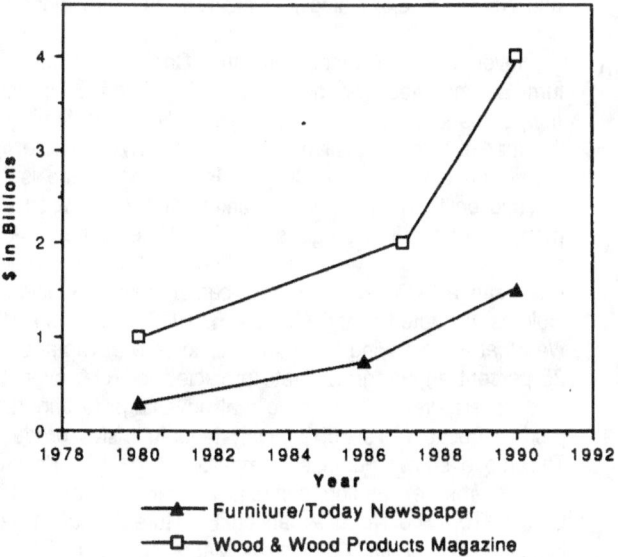

Figure 2.—*U.S. RTA industry sales volume estimates.*

The flexibility of design in RTA furniture could lead RTA, Inc., to expand into the kitchen cabinet market. The State of Minnesota study also found that 4 to 6 percent of kitchen cabinet sales in the United States are in RTA (this is a low share of the market, compared with Europe where up to 65 percent of cabinets are RTA and the United Kingdom where up to 75 percent are RTA). The 1986 U.S. retail market was $4.5 billion for 40.1 million cabinets in 5.6 million new or re-modeled kitchens; a stable annual growth of 1 to 3 percent has been documented. Increased sales of RTA kitchen cabinets have been to DIY customers at home improvement center stores, which account for three-fourths of the total of hardware and remodeling merchandise in the United States. The new 'frameless" or "European" design lends itself well to RTA and 32mm construction techniques.

The current trend is to incorporate more wood, solid and veneer, into the RTA products. Manufacturers must balance price and quality. Where a product would not sell for over $50 ten years ago, the break point is now $250 according to *Furniture/Today Newspaper.* The industry hopes for a $800 break point in 1990.

Paul Bush, President of Bush Industries, the fourth largest U.S. RTA furniture manufacturer as quoted in *Furniture/Today* says, "What we're making now is far better in style. It assembles better and is a better quality and better appearing type of product. There was at one time a lot of junk out there and that's what the industry has to overcome. Frankly, anybody who can upgrade and sell a higher-priced product is an asset to the industry because there's a real consumer need out there.

"We welcome innovative product, and hopefully ours as well as anybody elses will help the total industry," Bush said. "What we don't like to see is people out there that sell the stuff that isn't innovative. Those people are going to get weeded out and the consumer is not going to accept it."

RTA kitchen cabinets share all the advantages of other RTA furniture, such as reduced transportation costs, less shipping damage, less inventory space–and they cost roughly half the price of conventional cabinets. The Minnesota study states that the growing midwest market, now at $40 to $45 million for RTA kitchen cabinets, "could easily double within the next several years, however, as awareness and acceptance of RTA increases." While the study identified several RTA cabinet manufacturers, it could not find a major midwest-based producer.

B. Competitor Analysis

The main competition for a new U.S. manufacturer is imported RTA furniture from the Pacific Rim and Europe and some established U.S. manufacturers. Most design trends and improvements in construction come from Europe. Much imported RTA furniture, sold through discount store chains, mass merchandisers, and stores like IKEA, is the lowest quality and the lowest priced. Swedish-based IKEA, for example, invests in plants in countries of the world where the least-cost labor and resources are available; some plants are in the U.S.S.R. where IKEA is being paid back in product which they sell in their European and three U.S. stores. It would not be profitable to start a furniture manufacturing operation to compete with low-priced RTA furniture.

Some higher quality RTA furniture is also imported, and some is made domestically. In the Great Lakes area Marshall Erdman and Associates, Inc., located near Madison, Wisconsin, produced about $12 million worth of high-quality Techline furniture in 1986 and estimates that in 1988 their production will be at $18 million. High quality is achieved through use of the best available materials and machinery, maintenance of close tolerances, and controlling the finishing steps. In 1986 Marshall Erdman spent over $2 million for machinery to apply melamine to their particleboard as well as to produce coated board for other manufacturers.

Furniture/Today Newspaper states that 50 percent of the U.S. RTA furniture sales are produced by 5 domestic manufacturers. However, their estimate of the market size is about half of *Wood & Wood Products'* estimate that the current retail sales market is $2 billion. Table 1 shows *Furniture/Today's* summary of the "top 5 domestic producers" of RTA furniture as updated by *Furniture Design & Manufacturing's* "Top 300 North American Furniture Manufacturers" sales volumes.

Table 1.—

Summary of the top 5 domestic RTA furniture producers.

Company	Location	Total shipments (millions)	Estimated 1987 plant capacity (1000 square feet)
Foremost	Archibold, Ohio	$200	1,300
O'Sullivan	Lamar, Missouri	$135	1,000
Bush	Jamestown, New York	$65	576
Gusdorf	St.Louis, Missouri	$60	650
Charleswood	Chesterfield, Missouri	$55	600

Sources: *Furniture/Today Newspaper,* Volume 10, Number 45, August 18, 1986, page 23; *Furniture Design & Manufacturing,* Volume 60, Number 2, February 1988, pages 20-152.

Foremost Furniture, the largest domestic manufacturer and a divison of Sauder Woodworking Company, credits its success to their competitive production of low-priced entertainment centers, electronic and computer furniture, wall units, and bedroom furniture. Their competitors expect Foremost to remain the top U.S. manufacturer in terms of sales due to their reliable laminate finishing process and to their growing brand loyalty.

O'Sullivan Industries, Inc., also produces electronic and computer furniture, entertainment centers, and wall units as well as office furniture. O'Sullivan has moved from a low-priced general product line to veneered entertainment centers at higher prices. They have also entered a more conventional furniture market by selling some assembled entertainment centers.

Bush Industries, Inc., produces higher-priced electronic and office furniture, entertainment centers, and utility carts that include some solid wood. Bush is expanding their products to include more sophisticated and upscale products.

The oldest U.S. RTA furniture maker is Gusdorf Corporation which has been in business since 1946. They specialize in low-priced electronic and computer furniture, but also produce tables, utility carts, and office furniture. Gusdorf is experimenting with a broader line of products at higher price points in order to regain market share.

Charleswood manufactures a wide line of low- to medium-priced electronic, office, and bedroom furniture and wall systems. Charleswood distinguishes itself by offering products in the colors of white, black, rosewood, hickory, and maple.

The leading 5 RTA furniture manufacturers are discussed above, but other companies such as Affordable, Casard-Case, and Royal Creations are gaining in size. Some traditional furniture manufacturers like Singer, Thomasville, and Rowe are expanding into the RTA furniture market.

In November 1987 Singer Furniture Company, a conventional furniture manufacturer, purchased L&B Wood Specialties, Inc., a manufacturer of RTA furniture. Singer is familiar with the market through their past imports of RTA furniture from Brazil and the Far East. Singer's RTA furniture line is focused on the higher-priced market using solid wood in many products; it is Singer's belief that the future RTA products will be all wood. L&B Wood Specialties strength has been a $400 entertainment center. Singer President Jeff Holmes is quoted in the October 5, 1987 *Furniture/Today Newspaper* as saying, "We feel like to take advantage of the 25 percent annual growth rate projected for RTA over the next several years, there's no real advantage to importing the product because it's not a highly labor-intensive product." This move shows potential competition for RTA furniture makers from expanding conventional furniture manufacturers. Of the "Top 300 North American Furniture Manufacturers" identified by *Furniture Design & Manufacturing* in February 1988, 19 listed RTA furniture as one of their products; 8 of the 19 listed RTA furniture as their primary or only product.

With the low end flooded by imported RTA furniture and the high end dominated by established brand name manufacturers, the market has a share left for the middle-quality product. There will still be foreign and domestic competition for this market niche, but it is substantially less than competition at the extremes. The huge, growing demand for RTA furniture should strengthen the market position for medium-quality RTA product as consumers become educated about RTA furniture. With an initial orientation towards the medium-quality market, the manufacturer could eventually expand to a higher quality line.

C. Target Market and Positioning Strategy

While original KD furniture was inexpensive and poor quality and while it was purchased only by price shoppers, the trend towards improved design and quality means that medium-quality RTA furniture appeals to a wide range of consumers, a larger market. For those in the low-price market or KD lifestyle, the medium-quality product is a logical step up in price and quality. Conversely, the high-quality market chooses the less expensive, medium-quality RTA furniture because of its value. The market niche for the medium-quality product falls between the extremes and appeals to quality-conscious consumers looking for value in their purchases.

D. Marketing Mix Strategy

1. Product strategy.-Product strategy calls for producing a line of medium-quality RTA furniture consistent with the capabilities of the plant design. A keen marketing orientation, with designs and products that keep pace with innovations in the RTA furniture market, will maintain a progressive product. This market plan utilizes a single piece of furniture: a four-shelf bookcase, because bookcases, electronic and computer furniture, and storage units are the most popular pieces produced and sold (fig. 3). Growing demand for tables, chairs, desks, and kitchen cabinets could encourage and direct the expansion of the company.

2. Price strategy.-The price of medium-quality RTA furniture falls between prices for low-quality and high-quality RTA furniture. For example, a simple bookcase with 4 adjustable shelves that stands 50 3/4 inches tall by 27 1/2 inches wide and 11 3/4 inches deep (129 cm x 70 cm x 30 cm) with wood veneers on 5/8-inch particleboard would retail for $270 at the high end. At the low end, a comparable -'bookcase of 5/8-inch melamine coated particleboard retails for $75. A medium-quality manufacturer could easily produce to price within this range. While the example bookcase for RTA, Inc., uses oak-veneered particleboard and competes in quality with the top-priced furniture, the unit is priced at $70 wholesale (because this figure brought the gross and operating profits in line with data on similar manufacturing firms from Robert Morris Associates) and at $105 to $140 retail (a 50 percent to 100 percent mark-up).

3. Promotion strategy.—RTA, Inc., promotes its products through contact with RTA furniture buyers. Initially mail, phone, and direct sales calls introduce the furniture line to buyers. Furniture expositions such as the International

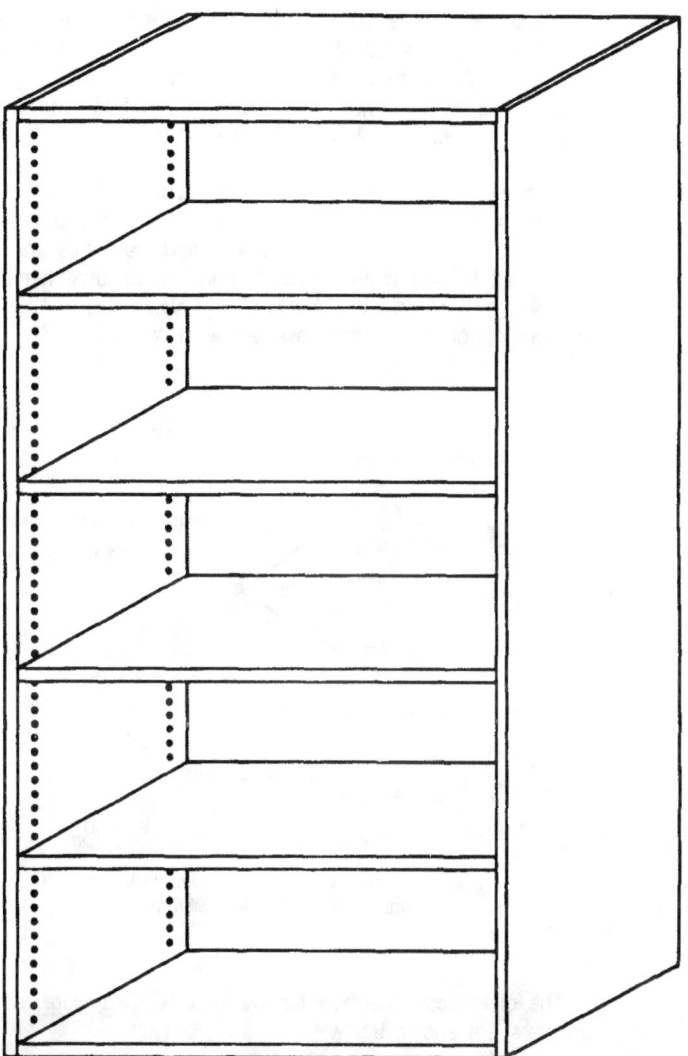

Figure 3.—*Open bookcase with 4 adjustable shelves.*

Ready-To-Assemble Furniture Show, the Cabinet Manufacturing and Marketing Seminar and Exposition, and the National Home Center Show (all held in the United States) are logical places to exhibit the new product line. Eventually the company will require permanent display showrooms at U.S. furniture marts where store buyers and interior designers select their purchases. Of course, a small display area within the plant should always be maintained too.

4. Distribution strategy.—The primary distribution paths for RTA furniture are mass merchandisers such as K-Mart, Sears, J. C. Penney, and Montgomery Wards and discount stores like Target, Caldor, and Gemco. Well-known retail chains, which often carry two or more lines of RTA furniture to segregate themselves from the mass merchandisers, are Storehouse, Workbench, Room & Board, and Conrans. Other RTA furniture outlets are contemporary specialty chains, including small lifestyle shops and Scandinavian stores importing RTA products direct from Europe. One of the newest and fastest growing markets for RTA furniture (30 percent increase from 1985 to 1986) is the home improvement stores; these stores often carry RTA wail cabinets, bathroom vanities, kitchen cabinets, and wall units. Increased RTA furniture sales through conventional furniture stores are forecast; *Furniture/Today Newspaper* breaks down current RTA furniture sales by store type (fig.4).

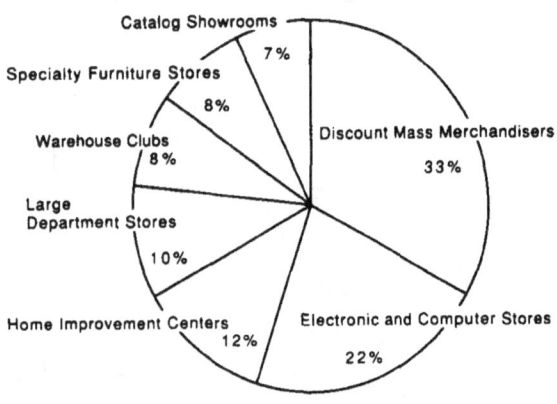

Figure 4. *RTA furniture retail distribution.*

Source: *Furniture/Today Newspaper,* Volume 10, Number 45, August 18, 1986, page 22.

The distribution channels for five U.S. RTA furniture manufacturers are shown below:

Company	Channels of distribution
Foremost	Discount mass merchants, large department stores, national chains, catalog showrooms, home centers, electronics specialty stores, furniture stores, and catalogs.
O'Sullivan	Electronics specialty stores, discount mass merchants, catalog showrooms, department stores.
Bush	Electronics specialty stores, discount mass merchants, department stores, catalog showrooms.
Gusdorf	Discount mass merchants, department stores, electronics specialty stores.
Charleswood	Discount mass merchants, home improvement centers, drug stores.

Source: *Furniture/Today Newspaper,* Volume 10, Number 45, August 18, 1986, page 23.

The National Home Furnishings Association's (NHFA) Operating Experiences Report for 1987 shows the following furniture type break down for their members' stores (fig. 5). Note that the NHFA membership does not include all the current distribution paths of foreign and domestic RTA furniture manufacturers (fig. 4).

Retailers see RTA furniture as a significant growth area and are allocating increasingly larger areas of floor space. Not only is floor space increasing, but many stores, like the 80-store Gemco chain in California, Arizona, and Nevada, include RTA furniture in their lifestyle departments located on the major aisles through which all customers pass to shop for their groceries. The distribution channels for RTA furniture are maturing, which will lead to an increased share of the furniture market.

K-Mart, which has 2,069 stores and over $21.5 billion in total sales in 1985, is the nation's largest discount department store and the second largest retail operation. Every store carries RTA furniture on 100 to 600 square feet of floor space. William Cashman, Senior Buyer for K-Mart, is expanded furniture departments in over 100 stores in 1986 to 4,000 to 5,000 square feet. According to a March 1986 *Furniture/Today* article, the discount retailers find their best-selling products are: wall units, room dividers and entertainment centers, priced to retail around $100; occasional chairs, ranging from folding chairs at $12.99 to lacquer dining chairs at $99; multi-functional foam seating/sleeping pieces available as sofas under the $200 retail, with correlating loveseats and chairs; occasional tables starting at $39.

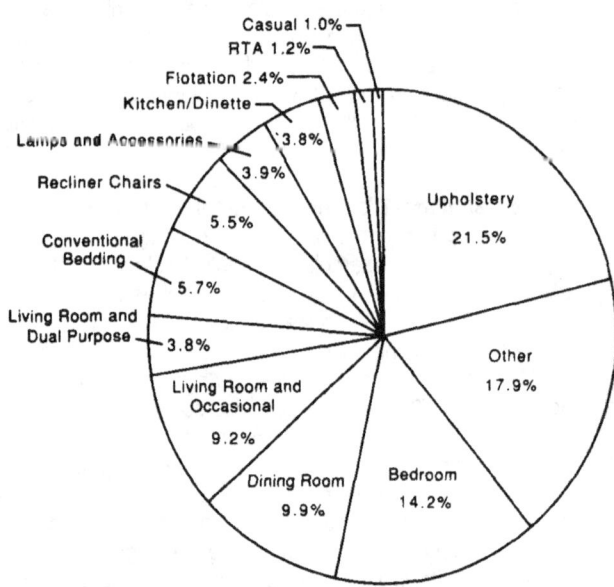

Figure 5.—*Types of furniture sold in NHFA members' stores.*

Source: *Furniture/Today Newspaper,* Volume 12, Number 5, October 5, 1987, page 1.

Executives at discount department stores (Target, Caldor, and Gemco) report that wood occasional furniture is the mainstay of their furniture business. Retailers agree that, in addition to basic occasional tables and wall units, the entertainment center is currently the fastest growing category.

At Caldor, which has 108 department stores that sell RTA furniture, the buyer said that 50 percent of the European furniture market is RTA, but only 2 percent of the domestic market is penetrated.

No discussion of RTA furniture retailers would be complete without documenting the newest arrival in the United States, IKEA. With 69 stores worldwide, the Swedish firm now has 3 stores in the United States. IKEA exemplifies a fairly novel method of merchandising: stores are supermarkets of furniture—all of the major pieces are RTA. In the first 5 days, IKEA drew 200,000 customers into its new Philadelphia store and had some 50,000 people inside at one point! The crowd was a result of intense pre-opening publicity that included giveaways of a car, trips to Sweden, and furniture. Every patron also got a wine glass, balloons, and other small promotional gifts. At Christmas time, IKEA sold Christmas trees for $5 which was refundable when the customer brought back the tree after Christmas; in the spring the customers could return with the receipt for the tree to pick up a couple of bushels of wood chips. With this strategy IKEA got the customer to visit the store three times!

IKEA's supermarket works by having the customer select purchases from the warehouse shelves and take them to the sales counter. The consumer's desire for immediacy is satisfied because purchases are taken home with the customer (a delivery service is available if desired). Bulkier items are picked up at the loading dock behind the store. An annual catalog is published and available to the customer for orders. IKEA's distinctive packages include a wrench, hardware, and instructions for assembly. IKEA offers contemporary kitchen, living, dining, and bedroom furniture.

They also have a wide selection of wall systems and home entertainment centers. Office furniture is available. In addition to furniture, IKEA also carries in their 15,000-item inventory, housewares, linens, toys, lamps, wallpaper, rugs, tiles, and fabric.

A May 1986 article in *Wood & Wood Products* states that the DIY market is now the fastest growing segment in the U.S. retail economy. Growth is projected at 8.6 percent for DIY versus 6.6 percent for overall retailing. Kitchen and bath remodeling are the largest and most profitable projects in the home improvement industry with a conservative estimate of $15 billion in annual sales; for example, new residential construction of 12 million kitchens and baths in 1986 came when only 1.8 million new houses were constructed. The NKCA expects 1988 gains in the repair and remodeling market, up 5.3 percent over 1987, will more than offset the 2.7 percent decline in the new residential construction market. In numbers, the strong economy will lead consumers to purchase 43.1 million cabinets in 1988. *Wood & Wood Products* feels the European influence, in terms of design and manufacture, will result in many of these cabinets being manufactured under the 32mm system and in RTA form. This DIY market can be reached primarily through the home improvement center stores.

Lowe's the largest home center chain in the nation installed 1,000 square foot RTA furniture departments in its 300 stores. Lowe's carries chairs, casual dining, occasional tables, wall systems, and electronic furniture according to *Furniture/Today Newspaper* (November 10, 1986).

5. Sales strategy.—The marketing and sales manager work full time to direct the marketing functions such as promotions, market development, contacts with buyers, and sales. The manager continually assesses the market for new designs, trends, construction methods, fasteners, and prices of competitors.

A. Organizational Form

The fledgling firm, RTA, Inc., incorporates as a small business known as an "S corporation." The regulations of the particular state in which the corporation locates affect the exact structure and taxation. The advantages of the S corporation are limited liability, recognition as a legal entity, favorable tax treatment, relative ease of transferability of interests, and continuity.

B. Management Team and Staff

The management team includes the general production manager and the sales manager. These two individuals are the basic corporation leaders and are intrinsically involved in RTA, Inc.'s startup and daily operations. Eventually the management team could expand with the corporate growth.

The remaining staff consists of one production foreman, one office manager who is a secretary-bookkeeper, one panel saw operator, one panel saw tailer, one edgebander operator, one edgebander tailer, one boring machine operator, one person for the finishing operation, two people in packaging and shipping, and one person to drive the forklift removing scrap and to do odd jobs facilitating production. In addition to the 2 managers, 11 people are needed; funding to automate some machines and transfers between stations would reduce the number of workers.

Most likely, the general production manager and the sales manager also function as the corporate officers of president and secretary-treasurer, unless there is a parent company or private investors. The corporation should explore the availability of qualified people to form a board of directors that actually helps run the company.

Supporting professional services of a certified public accountant, legal counsel, financial advisor, banker, and a business consultant should be established early in the planning process.

C. Ownership

Ownership of RTA, Inc., depends on the organization of the corporation, such as sole proprietor, partnership, or group of investors. There are many ways that public and private stock could be distributed. While unlikely, it is possible that local government's or a local economic development group's financing would affect the ownership too.

A. Location

Locating this plant in the Northeastern Area places the products within one to two days' trucking distance of over one-half the U.S. population. The state incentives available to new manufacturers vary, as seen in "Industrial Development Incentives in Northeastern Area States" in Appendix A. When the firm has narrowed its choice of location, it should contact each state's Department of Industrial Development (or its equivalent) for details of the incentives listed in the table and information about any additional incentives. Most states utilize the federal industrial development incentives (which are not listed in the table) such as industrial revenue bonds and worker training.

B. Labor Force and Labor Costs

Listed below are the corporation's entire labor force, including managers and staff, and the labor costs of their salaries and fringe benefits.

Position	Annual salary or wages
General production manager (corporate president)	$30,000
Sales manager (corporate secretary-treasurer)	30,000
Office manager and secretary-bookkeeper	20,000
Production foreman	25,000
Panel saw operator	20,000
Panel saw tailer	20,000
Edgebander operator	20,000
Edgebander tailer	20,000
Boring machine operator	20,000
Finishing person	20,000
Packaging and shipping person	15,000
Packaging and shipping person	15,000
Forklift driver and odd-job person	15,000
Total with 20% fringe benefits added	$324,000

C. Materials Procurement

Prices for anticipated inventory needs were collected in order to project values for the financial plan, Part 6 of this report. It suffices to say that finished and unfinished board is available and that contracts for quantity orders cut material costs according to volumes purchased. Individual costs may be found in the "Notes" included in Section E "Statement of Projected Operations and Cash Flows" of Part 6. Shipments of all incoming materials and finished goods are by commercial truck.

D. Production Schedule

At the start of production (and until the plant is running smoothly) one shift will run. We expect only about 50 percent of production capacity the first year; in the second year more efficient utilization of plant and equipment can be achieved by adding a second full shift to double production.

Year 1: 1 shift at 40 hours/week, 8 hours/day

40 hours/week x 50 weeks/year = 2,000 hours/year

Year 2: 2 shifts at 80 hours/week, 16 hours/day

80 hours/week x 50 weeks/year = 4,000 hours/year

Production could eventually be more than tripled by using a "rolling 40" work week. Use of the rolling 40 production schedule is controversial in highly unionized areas, but has good acceptance in right-to-work states. The difference in work hours during the first and second years is illustrated below. A more detailed description of the rolling 40 with 4 shifts appears in Appendix B.

E. Long-range Plans

The ultimate goal for the investors is to build a company that will be salable when they wish to leave the business. To accomplish this goal, the company must be progressive in its designs, production techniques, marketing strategies, and business operation.

The company should plan to grow from the medium-quality RTA furniture market to the high-quality RTA furniture market because profit margins will be greater at the upper end. If experience in marketing and producing medium-quality RTA furniture leads the firm to establish a favorable reputation, the firm should have few problems entering the high-quality RTA furniture market. A progressive company continually seeks profitable opportunities for expansion, such as production of kitchen and bathroom cabinets compatible with the company's production capabilities.

Part 6
FINANCIAL PLAN

A. Capital Requirements
The capital requirements of RTA, Inc., for the first year are below:

Machinery (panel saw, edgebander, boring machine, and finishing equipment)	$55,000

Working capital

12 months' salaries and wages	324,000
Contingency money	100,000
Receivables	100,000

Inventory for 26,100 units (100 units/working day)

Particleboard: oak-veneered, 5/8-inch thick, 45-pound, 4-foot by 8-foot (1 /unit x $23/board)	600,300
Fasteners and hardware (36/unit x $0.10/piece)	93,960
Edgebanding (80 feet/unit x $0.018/lineal foot)	37,584
Catalyzed varnish (128 square feet/unit for 2 coats x $0.0257812/square foot)	86,130
Shipping boxes (1 box/unit x $2.10/box)	54,810
Pallets for shipping ($4/pallet x 2/day)	2,088
Total working capital	1,398,872
Forklift and 20 factory trucks	12,000
Miscellaneous tools ($2,000) plus factory supplies ($5,000)	7,000
Office furniture and equipment ($5,000) plus supplies ($1,200)	6,200
Total	$1,479,072

B. Financing Plan
Financial grants and low interest loans are available from the federal government and from most state or local governments (see "Industrial Development Incentives" in Appendix A).

Some states have operating capital loans that will be necessary in the second year when production is increased. Often state and local governments and economic development associations will provide land, site improvements, and buildings to entice new businesses to locate in their area (these industrial development incentives would take the place of rent in the "Statement of Projected Operations and Cash Flows for the First 12 Months" in Section E). Regardless of the sources of financing, grants, government loans, private bank loans, or public and private stock, some proportion of financing (10 to 25 percent) should be borne by the investors.

C. Beginning Balance Sheet
A pro forma balance sheet is offered here only to show the format and the entries for the scenario described thus far.

RTA, Inc.
Pro Forma Balance Sheet
First Year of Operation

Assets		
Current assets		
Cash	$524,000	
Inventory	874,872	
Supplies	6,200	
Fixed assets		
Machinery	55,000	
Forklift, factory carts	12,000	
Miscellaneous tools	2,000	
Office equipment	5,000	
Total assets		$1,479,072
Liabilities and owners' equity		
Current liabilities (year 1 salaries and wages)	324,000	
Long-term liabilities (machinery)	55,000	
Owners' equity	1,100,072	
Total liabilities and owners' equity		$1,479,072

D. Statement of Projected Operations and Cash Flows for the First 12 Months

(See next page)

STATEMENT OF PROJECTED OPERATIONS AND CASH FLOWS FOR THE FIRST 12 MONTHS *(in dollars)*

SALES PLAN

Month in Year 1 of Operation	January	February	March	April	May	June
UNIT SALES PLANNED	– 0 –	– 0 –	2,300	2,100	2,200	2,300
DOLLAR SALES PLANNED	– 0 –	– 0 –	161,000	147,000	154,000	161,000

FINANCIAL PLAN

	January	February	March	April	May	June
SALES	– 0 –	– 0 –	161,000	147,000	154,000	161,000
COST OF GOODS SOLD						
Materials						
Veneered particleboard[1]	– 0 –	48,300	52,900	48,300	50,600	52,900
Fasteners and hardware[2]	– 0 –	7,560	8,280	7,560	7,920	8,280
Edgebanding[3]	– 0 –	3,024	3,312	3,024	3,168	3,312
Finishing[4]	– 0 –	6,930	7,590	6,930	7,260	7,590
Shipping boxes and pallets[5]	– 0 –	4,578	5,014	4,578	4,796	5,014
Total materials	– 0 –	70,392	77,096	70,392	73,744	77,096
Direct labor[6]	– 0 –	12,075	13,225	12,075	12,650	13,225
Overhead						
Indirect labor[7]	– 0 –	3,213	3,519	3,213	3,366	3,519
Benefits[8]	– 0 –	3,058	3,349	3,058	3,203	3,349
Equipment depreciation[9]	– 0 –	983	983	983	983	983
Utilities and supplies[10]	– 0 –	2,500	2,500	2,500	2,500	2,500
Total overhead	– 0 –	9,754	10,351	9,754	10,052	10,351
Total cost of goods sold	– 0 –	92,221	100,672	92,221	96,446	100,672
GROSS PROFIT	– 0 –	-92,221	60,328	54,779	57,554	60,328
SELLING, GENERAL, AND ADMINISTRATIVE EXPENSE						
Salaries[11]	6,667	6,667	6,667	6,667	6,667	6,667
Benefits[8]	1,333	1,333	1,333	1,333	1,333	1,333
Freight[12]	– 0 –	– 0 –	4,830	4,410	4,620	4,830
Travel[13]	– 0 –	1,800	1,800	1,800	1,800	1,800
Rent[14]	4,000	4,000	4,000	4,000	4,000	4,000
Office supplies[15]	100	100	100	100	100	100
Depreciation[16]	75	75	75	75	75	75
Miscellaneous S,G,&A expense[17]	100	500	5,000	500	500	500
Total S,G,&A expense	12,675	14,475	23,805	18,885	19,095	19,305
OPERATING PROFIT	-12,675	-106,696	36,523	35,894	38,458	41,023
OTHER INCOME (Expense)						
Interest expense[18]	– 0 –	– 0 –	– 0 –	– 0 –	– 0 –	– 0 –
Net other income	– 0 –	– 0 –	– 0 –	– 0 –	– 0 –	– 0 –
EARNINGS BEFORE TAXES	-12,675	-106,696	36,523	35,894	38,458	41,023
RESULTS OF OPERATIONS						
Earnings before taxes	-12,675	-106,696	36,523	35,894	38,458	41,023
Depreciation	– 0 –	917	917	917	917	917
Cash available from operations	-12,675	-105,779	37,440	36,811	39,375	41,940
CASH PROVIDED BY:						
Increase in accounts payable[19]	88,738	8,451	-8,451	4,226	4,226	-4,226
Total cash provided	88,738	8,451	-8,451	4,226	4,226	-4,226
TOTAL CASH AVAILABLE	76,062	-97,328	28,989	41,037	43,601	37,714
CASH USED BY:						
Increase in inventory[20]	50,000	-8,451	4,226	4,226	-4,226	4,226
Increase in accounts receivable[21]	– 0 –	– 0 –	– 0 –	241,500	-21,000	10,500
Decrease in accounts payable	– 0 –	– 0 –	– 0 –	– 0 –	– 0 –	– 0 –
Paydown of loans[18]	– 0 –	– 0 –	– 0 –	– 0 –	– 0 –	– 0 –
Purchase of equipment[22]	74,000	– 0 –	– 0 –	– 0 –	– 0 –	– 0 –
Total cash used	124,000	-8,451	4,226	245,726	-25,226	14,726
CASH EXCESS (Shortfall)	-47.938	-88,877	24,763	-204,689	68,827	22,989
CUMULATIVE CASH POSITION	-47,938	-136,814	-112,051	-316,740	-247,914	-224,925

**See notes on next page

July	August	September	October	November	December	TOTALS
2,200	2,300	2,200	2,100	2,100	2,200	22,300
154,000	161,000	154,000	147,000	147,000	154,000	1,540,000
154,000	161,000	154,000	147,000	147,000	154,000	1,540,000
50,600	52,900	50,600	48,300	48,300	50,600	554,300
7,920	8,280	7,920	7,560	7,560	7,920	86,760
3,168	3,312	3,168	3,024	3,024	3,168	34,704
7,260	7,590	7,260	6,930	6,930	7,260	79,530
4,796	5,014	4,796	4,578	4,578	4,796	52,538
73,744	77,096	73,744	70,392	70,392	73,744	807,832
12,650	13,225	12,650	12,075	12,075	12,650	138,575
3,366	3,519	3,366	3,192	3,213	3,366	36,852
3,203	3,349	3,203	3,053	3,058	3,203	35,085
983	983	983	983	983	983	10,813
2,500	2,500	2,500	2,500	2,500	2,500	27,500
10,052	10,351	10,052	9,728	9,754	10,052	110,250
96,446	100,672	96,446	92,195	92,221	96,446	1,056,657
57,554	60,328	57,554	54,805	54,779	57,554	483,343
6,667	6,667	6,667	6,667	6,667	6,667	80,004
1,333	1,333	1,333	1,333	1,333	1,333	16,001
4,620	4,830	4,620	4,410	4,410	4,620	46,200
1,800	1,800	1,800	1,800	1,800	1,800	19,800
4,000	4,000	4,000	4,000	4,000	4,000	48,000
100	100	100	100	100	100	1,200
75	75	75	75	75	75	900
500	500	500	500	500	500	10,500
19,095	19,305	19,095	18,885	18,885	19,095	222,605
38,458	41,023	38,458	35,919	35,894	38,458	260,738
– 0 –	– 0 –	– 0 –	– 0 –	– 0 –	– 0 –	– 0 –
– 0 –	– 0 –	– 0 –	– 0 –	– 0 –	– 0 –	– 0 –
38,458	41,023	38,458	35,919	35,894	38,458	260,738
38,458	41,023	38,458	35,919	35,894	38,458	N/A
917	917	917	917	917	917	N/A
39,375	41,940	39,375	36,836	36,811	39,375	N/A
4,226	-4,226	-4,251	25	4,226	-4,226	N/A
4,226	-4,226	-4,251	25	4,226	-4,226	N/A
43,601	37,714	35,125	36,861	41,037	35,150	N/A
-4,226	-4,251	25	4,226	-4,226	-4,394	N/A
10,500	-10,500	10,500	-10,500	-10,500	– 0 –	N/A
– 0 –	– 0 –	– 0 –	– 0 –	– 0 –	– 0 –	N/A
– 0 –	– 0 –	– 0 –	– 0 –	– 0 –	– 0 –	N/A
– 0 –	– 0 –	– 0 –	– 0 –	– 0 –	– 0 –	N/A
6,274	-14,751	10,525	-6,274	-14,726	-4,394	N/A
37,327	52,465	24,599	43,136	55,762	39,543	N/A
-187,599	-135,134	-110,534	-67,398	-11,636	27,907	N/A

Notes:

[1] Oak-veneered, 5/8-inch thick, 45-pound, 4-foot-by-8-foot particleboard or medium density fiberboard at $23 per board, one board per unit.

[2] 36 fasteners and pieces of hardware per unit at $0.10 each or $3.60 per unit.

[3] Estimated at 80 feet per unit at $0.018 per lineal foot or $1.44 per unit.

[4] Estimated at $3.30 per unit: $12.36 per gallon catalyzed conversion varnish with 480 square foot coverage (23% overspray allowance); 64 square feet per unit, 2 coats varnish required, 0.13 gallon per coat.

[5] Printed shipping boxes and padding at $2.10 per box; one box per bookcase. Pallets at $4 each, 2 pallets per one day's production of 100 units, $0.08 per unit. Combined cost, $2.18 per unit.

[6] Includes only machinery operators and tailers and combination shipper-and-packagers.

[7] Includes plant foreman and combination forklift and odd-job person.

[8] Include workers' compensation, payroll taxes and benefits, 20% of total salaries and labor.

[9] Calculated using the conservative straightline depreciation method; salvage value at the end of the 5-year depreciation period estimated to be $10,000 on the main equipment, forklift, factory trucks, and handtools combined.

[10] Utilities, mainly electricity and heat, $3,000 per month for 2 shifts or in year 1, $2,000 per month for one shift; factory supplies, $500 per month.

[11] Include general production manager, sales manager, and office manager/secretary-bookkeeper. Eventually commissions might be paid to manufacturers' representatives who help sell the line of products, but commissions are not included in this analysis.

[12] Assumed to be 3% of sales.

[13] Year 1, sales manager travels extensively to generate orders and attend trade shows, 12 travel days per month at $150 per day.

[14] Includes $1,500 per month for 5,000 square feet, sufficient for the machinery and required surge areas, shipping, storage of incoming materials and finished inventory, and office space.

[15] Estimated at $200 per month.

[16] Office furniture and equipment for 3 people, estimated at $5,000, straightline depreciated over 5 years to minimal salvage value $500; therefore, depreciation is $900 per year or $75 per month.

[17] Includes insurance as well as other miscellaneous selling, general, and administrative expenses such as printing product literature, advertising, promotional items, and other smaller selling expenses. March $5,000 promotional cost covers major items for the year.

[18] No allowance because state or federal grant is assumed as is owners' equity in the business. However, this item appears in the statement to draw attention to interest expenses recording location if a loan is taken out and included in the records.

[19] Results from cost change from one month to the next month for manufacturing and direct labor plus material costs of board, fasteners, edgebanding, and packaging: also known as "materials purchased."

[20] Equals the difference in total cost of goods sold in the month after next and the next month. For example, July's increase in inventory is September's $96,446 - August's $100,672 = ($4,226).

[21] Equals 45 days of previous month's average sales level minus the previous month's increase in accounts receivable balances. For example, July increase in accounts receivable is $161,000 x 1.5 months - June's $10,500 - May's ($21,000) - April's $241,500 = $10,500.

[22] Totals $57,000, including $51,000 for panel saw, edgebander, and boring machine, plus $4,000 for finishing equipment with spraying equipment and spray booth, plus $2,000 in miscellaneous tools.

E. Investment Criteria

The EVALUE computer program is used in order to calculate the investment criteria of internal rate of return, payback period, present net worth, and present net worth to initial investment ratio. The complete printout for EVALUE, the microcomputer program for determining the financial feasibility of forest products industry investments, appears in Appendix C. After reproducing the assumptions input into the program, EVALUE produces these statements: Revenues and Costs, Working Capital, Initial Investment and Salvage Value, Depreciation, Profits and Earnings and Cash Flows, and Discounted Cash Flows and Investment Criteria. Section D gives a statement of projected operations and cash flows for RTA, Inc's first year but EVALUE has a 5-year time horizon. Another difference between these two investment examination scenarios is that actual production does not begin until the third month in the Statement of Projected Operations and Cash Flows, where in EVALUE, as required by the computer program, costs were scaled up to reflect a full 12 months of operation for the first year.

The EVALUE computer program assumptions are below:

Length of planning period	5 years
First year production (2,000 hours in year 1 / 4,000 hours in year 2)	50%
Annual inflation rate	5%
Depreciation (straightline schedule for 5 years on machinery, forklift, carts, tools, office equipment)	$74,000
Initial working capital (salaries $324,000 + contingency money $100,000 + inventory $874,872 + receivables $100,000)	$1,398,872
Annual gross revenues (year 1 revenues x 2 shifts in year 2)	$3,654,000
Raw material costs ($874,872 inventory x 2 shifts)	$1,749,744
Other annual variable costs (utilities $30,000 + freight $54,810 + labor $228,858 x 2 shifts)	$627,336
Annual fixed costs (salaries $96,000 + travel $21,600 + rent $48,000 + supplies $1,200)	$166,800
Tax rate	47%
Salvage value of property in year 5	$10,500
Discount or current interest rate.	9.5%

The EVALUE summaries of the cash flow show a payback period of 3.2 years based on after-tax net flow. The internal rate of return based on after-tax net cash flow is 33.7 percent, and the present net worth/initial investment ratio is 110.5 percent. The present net worth at the 9.5 percent discount rate is $1,626,832.

Part 7
RISKS AND WEAKNESSES

Obviously, this business plan is based on the best assumptions applicable at the date of writing. Changes in the assumptions will be necessary to customize the business plan for the investors, the production process or capabilities, the products produced, and the current economic climate. When given a choice in performing this analysis, conservative values were used.

Should there be a downturn in the U.S. economy, the market for RTA furniture could soften. The cost of materials in board, edgebanding, and hardware could rise—especially if the firm manufactures other products in addition to the simple bookcase used as an example in this business plan. The profit margin may decrease if the RTA furniture manufacturer becomes locked into a mass merchandiser that cuts the wholesale prices paid for the finished products.

This plan could change considerably, depending on the desires and experience of the investors-but it provides a basis for starting a customized plan.

APPENDIX

APPENDIX A
INDUSTRIAL DEVELOPMENT INCENTIVES IN NORTHEASTERN AREA STATES

State	INCENTIVES[1]					
	Financial grants	Low interest loans	Reduced state tax	Reduced local tax	Land & building grants[2]	Others? See notes
Connecticut	Y[3]	Y	Y	Y	Y	4
Delaware	Y	N	Y	Y	Y	5
Illinois	N	Y	Y	Y	Y	—
Indiana	N	Y	N	Y	Y	—
Iowa	N	N	Y	Y	N	—
Maine	N	Y[6]	N	Y	Y	4, 7
Maryland	Y[8]	Y[6]	Y[9]	Y	Y	4
Mass.	N	Y	N	N	Y	—
Michigan	Y	Y	N[10]	Y	Y	4, 11
Minnesota	Y[12]	Y	Y	Y	Y	4
Missouri	Y[8]	Y	Y	Y	Y	13
New Hamp.	N	N[5]	N	Y	N	—
New Jersey	Y[14]	Y	Y	Y	Y	4
New York	N	Y	Y	Y	Y	4
Ohio	Y	Y[15]	Y	Y	Y	—
Pennsylvania	N	Y[16]	N	Y	Y	17
Rhode Island	N	Y	N	N	Y	—
Vermont	N	Y[18]	N	Y	Y	—
W. Virginia	N	Y[19]	Y[20]	Y	Y	4
Wisconsin	N	N	N	N	Y	—

[1] In addition to the incentives listed, each state's forest products utilization department is capable of providing information on wood resources, supply sources, markets for wood and wood products, and the current wood products industry within the state. Federal government sponsored programs that are universally used by the states such as industrial revenue bonds, federal loan programs, and worker training programs are not included in this table.

[2] Often arranged through local community governments.

[3] $1000 per job created in Urban Enterprise Zones.

[4] A variety of additional assistance programs available.

[5] Tax exemption for export trading and foreign sales office; favorable trust, legal, incorporation, inheritance, and tax system.

[6] Guarantees bank loans.

[7] Central Maine Power keeps lists of available vacant buildings.

[8] Community Development Block Grant available through local community governments.

[9] Enterprise zones of high unemployment with tax advantages; free foreign trade zones, no duty until shipped.

[10] Capital investment tax credit available.

[11] Guaranteed loans, venture capital, no inventory or pollution control equipment taxes.

[12] Grants available through local community governments.

[13] "Rural Missouri" program with loans to wood products companies.

[14] Scarce.

[15] Up to $1 million at 3% to 4% interest rate; money from state liquor tax.

[16] 3% to 6% interest rate.

[17] Revolving loan fund for working capital; low interest.

[18] $200,000 maximum, 4% interest rate.

[19] 4% interest rate.

[20] Sliding scale; amount of reduction depends on number of jobs created.

Source: Discussion with Northeastern Area states' specialists in forest products utilization and economic development, December, 1986.

APPENDIX B
OPTIONAL ROLLING 40 WORK SCHEDULE

Option: 4 shifts using rolling 40 production schedule, 10 hours/day

10 hours/day/shift x 2 shifts/day = 20 hours/day

20 hours/day x 355 days/year = 7,100 hours/year

	Shift			
	1	2	3	4
	Day	Night	Day	Night
Work	M,T,W,R	M,T,W,R	F,S,Su,M	F,S,Su,M
Off	F,S,Su,M	F,S,Su,M	T,W,R,F	T,W,R,F
Work	T,W,R,F	T,W,R,F	S,Su,M,T	S,Su,M,T
Off	S,Su,M,T	S,Su,M,T	W,R,F,S	W,R,F,S
	et cetera	et cetera	et cetera	et cetera

Key: M=Monday, T=Tuesday, W=Wednesday, R=Thursday,
F=Friday, S=Saturday, Su=Sunday

APPENDIX C
EVALUE COMPUTER PROGRAM PRINTOUT

RTA, INC.

ASSUMPTIONS

Length of planning period	5 Years
Year 1 of planning period	1988
Startup level	50%
Inflation factor	5%
Nondepreciable property	$0
3-Year depreciable property	$0
5-Year depreciable property	$74,000
15-Year depreciable property	$0
Working capital	$1,398,872
Gross revenues	$3,654,000
Raw material costs	$1,749,744
Other variable costs	$627,336
Fixed costs	$166,800
Depreciation schedule	Straightline
Tax rate	47%
Salvage value	$10,500
Discount rate	9.5%

WORKING CAPITAL
(Inflation factor: 5%)

Year	Beginning	Added
1	$1,398,872	$69,944
2	1,468,816	73,441
3	1,542,256	77,113
4	1,619,369	80,968
5	1,700,338	85,017

DEPRECIATION
(Straightline depreciation)

Year 1	$14,800
Year 2	14,800
Year 3	14,800
Year 4	14,800
Year 5	14,800

(Continued on next page)

22

REVENUES AND COSTS
(Startup level: 50%, Inflation factor: 5%)

Year	Gross revenues	Raw material costs	Other variable costs	Fixed costs
1	$1,827,000	$874,872	$313,668	$166,800
2	3,836,700	1,837,231	658,703	175,140
3	4,028,535	1,929,093	691,638	183,897
4	4,229,962	2,025,547	726,220	193,092
5	4,441,460	2,126,825	762,531	202,746

PROFITS, EARNINGS, AND CASH FLOWS

Year	Profit before taxes	After-tax profit	After-tax earnings	After-tax net cash flow
0	$0	$0	$0	$1,472,873
1	471,660	249,536	264,336	194,392
2	1,165,626	609,938	624,738	551,297
3	1,223,907	640,827	655,627	578,514
4	1,285,103	673,260	688,060	607,092
5	1,349,358	707,316	2,517,970	2,517,970

INITIAL INVESTMENT AND SALVAGE VALUE

initial investment

Nondepreciable property	$0
3-Year depreciable property	0
5-Year depreciable property	74,000
15-Year depreciable property	1,398,872
Total	$1,472,872

Salvage value and recovered working capital at end of year 5

Salvage value	$10,500
Recovered working capital	1,785,355
Total	$1,795,855

DISCOUNTED CASH FLOWS AND INVESTMENT CRITERIA

Present net worth at 9.5% discount	$1,626,832
Payback period based on after-tax net cash flow	3.24 Years
Internal rate of return based on after-tax net cash flow	33.70%
Present net worth/initial investment ratio	110.45%

Appendix C

Important Formulas Which Correspond to the Headings in the Detailed Example In Appendix B

- Gross Profit=

  Sales minus cost of goods sold.

- Operating Profit=

  Gross profit minus selling, general, and administrative expense.

- Earnings Before Taxes=

  Operating profit plus net other income (expense).

- Cash Available From Operations=

  Earnings before taxes plus depreciation.

- Increase in accounts payable=

  Total cost of goods sold for the next month.

- Total Cash Available=

  Cash available from operations plus total cash provided.

- Increase In Inventory=

  Total cost of goods sold (COGS) from 2 months hence minus total COGS for next month.

- Increase In Accounts Receivable=

  45 days of previous month's average sales level minus the previous month's increase in accounts receivable balances.

- Cash Excess (shortfall)=

  Total cash available minus total cash used.

- Cumulative Cash Position=

  Each preceding period's cash excess or shortfall plus the current period's cash excess or shortfall.

For additional information or copies, please contact:

Stephen M. Bratkovich
Forest Products Specialist
Northeastern Area, State & Private Forestry
USDA Forest Service
1992 Folwell Avenue
St. Paul, MN 55108
Phone: 612-649-5246
FAX: 612-649-5238

Jeffrey L. Howe
Research Assistant
Forest Products Management
Development Institute
Department of Forest Products
University of Minnesota
2004 Folwell Avenue
St. Paul, MN 55108
Phone: 612-625-5200
FAX: 612-625-6286

✪U.S. GOVERNMENT PRINTING OFFICE: 1995-659-222/20067

www.ingramcontent.com/pod-product-compliance
Lightning Source LLC
Chambersburg PA
CBHW081600170526
45166CB00009B/2769